Machine Translation

Machine Translation

How Far Can It Go?

Makoto Nagao
Department of Engineering,
Kyoto University,
Kyoto, Japan

translated by
Norman D. Cook
Department of Experimental Psychology,
University of Oxford

Oxford   New York   Tokyo
OXFORD UNIVERSITY PRESS
1989

Oxford University Press, Walton Street, Oxford OX2 6DP
Oxford New York Toronto
Delhi Bombay Calcutta Madras Karachi
Petaling Jaya Singapore Hong Kong Tokyo
Nairobi Dar es Salaam Cape Town
Melbourne Auckland
and associated companies in
Berlin Ibadan

Oxford is a trade mark of Oxford University Press

Published in the United States
by Oxford University Press, New York

KIKAI HON'YAKU WA DOKO MADE KANO KA
by Makoto Nagao
Copyright © 1986 by Makoto Nagao
Originally published in Japanese by
Iwanami Shoten, Publishers, Tokyo, 1986
This translation © Norman Cook, 1989

British Library Cataloguing in Publication Data
Nagao, Makoto, 1936–
Machine translation how far can it go?
1. Languages. Translation. Applications
of computer systems
I. Title II. Kikai hon'yaku wa doko
made kano ka. English
418'.02
ISBN 0–19–853739–5

Library of Congress Cataloging in Publication Data
Nagao, Makoto, 1936–
[Kikai hon'yaku wa doko made kano ka. English]
Machine translation : how far can it go? / Makoto Nagao ;
translated by Norman D. Cook.
p. cm.
Translation of: Kikai hon'yaku wa doko made kano ka.
1. Machine translating. I. Title.
P308.N2613 1989 418'.02—dc19 88–19949
ISBN 0–19–853739–5

Set by Graphicraft Typesetters Limited
Printed in Great Britain
at the University Printing House, Oxford
by David Stanford
Printer to the University

Translator's preface

This book is a translation of *KIKAI HON'YAKU WA DOKO MADE KANO KA* by Makoto Nagao, Professor of Engineering at Kyoto University and one of Japan's leading experts in the field of machine translation.

The translation of this work was commissioned by Oxford University Press, partly because of the paucity of overviews on machine translation and partly because of the unusual perspective, from a Western point of view, of Japanese work in this area. Not only is Japanese research in this and directly related fields growing, both as part of the so-called Fifth Generation Computer Project and outside it, but the magnitude of the dissimilarities between English and Japanese makes the Japanese perspective on translation a particularly interesting one.

In a world where microcomputers are everywhere and writing with a wordprocessor is commonplace, the mere thought of machine translation — that is, translation from one natural language to another by means of computer — still strikes most people as a part of science fiction. Indeed, unless one has made an effort to track down a machine translation system, probably at a research institute, it is unlikely that one will have come across a working system or a machine-translated text.

As Professor Nagao points out, this situation is likely to change greatly throughout the world in the coming years, for a combination of reasons. Foremost among these is the increased availability of relatively inexpensive computers which contain a large enough memory for massive bilingual dictionaries to be stored within them. Secondly, the software, which has to deal with the subtleties and complexities of natural languages, is beginning to reach a level where significant things can be accomplished. Thirdly, there is a continuing and, in some quarters, a growing need to achieve automated means of producing intelligible (if perhaps imperfect) translations at a reasonable cost.

Both the hardware and software factors are a function of

international conditions: changes in the market-place for micro-chips and academic developments in linguistic theory. The perceived need for machine translation, however, is very unevenly distributed around the globe. In the English-speaking world, where much of the progress in linguistics has originated, there has been relatively little need to develop machine translation systems. In most of the Third World, where inexpensive translation would undoubtedly be welcomed, the costs of research and development are prohibitive. Work on machine translation has continued at several locations in Europe, but Japan is unusual in that the need for machine translation is perceived to be high and the resources required for research and development are not prohibitively large. Japanese research is therefore very much at the forefront of current developments in machine translation. At the same time, however, the language barrier — the existence of which is the motivating force behind research in mechanical translation — has prevented Japanese efforts in this direction from becoming widely known in the West. This volume addresses this problem and is a succinct, non-technical description of current work in progress there.

For those native speakers of English who gained some familiarity with French or Spanish during the early years of education, the prospect of machine translation (at least among the Indo-European languages, and particularly those with Latin at their roots) may appear trivial or, at best, uninteresting. The novice may think that a big enough electronic dictionary, plus a list of colloquialisms and perhaps a table with the irregular verbs, would more or less suffice to allow a computer to do a reasonably good job of translating factual text.

Machine translation has in fact proved to be a good deal more complex than that optimistic outlook would suggest, even for languages which have common historical roots. Inevitably, quite complex analysis of the grammatical structure of sentences and at least some information about the meanings of the words involved is required. In the case of translation between Japanese and English, the depth of analysis and extent of structural changes required for successful translation are evident from the outset.

The difficulties in translating between Japanese and English begin with the completely different writing systems, but simple transliteration from one system to the other reveals the depth of the differences that remain. A written Japanese sentence will typically contain phonemic characters, the *kana*, mixed together with the imported Chinese characters, the *kanji*. Romanization will then make the words accessible, but the conveniently placed spaces between words used in English are not used in Japanese, sothattheromanizedsentencewilllooklikethis. Only familiarity with the language will allow the sentence to be broken into its word units, but it is found that the transliteration will then have lost much of its original clarity due to the large number of words in Japanese which sound like one another but are written with different characters (homophones). My dictionary lists fourteen words pronounced as 'KANSHO' and having meanings as distinct as 'dishonest merchant' and 'intuition'.

Even when this problem has been surmounted, there remain many fundamental differences between English and Japanese that point to the unlikelihood of successful literal, word-for-word translation. In Japanese, plurals are only sometimes indicated for nouns, and never for verbs or adjectives. 'Adjuncts' which follow noun phrases are used to indicate the relation of nouns to verbs, but they are used in a less rigid manner than the unforgiving suffixes of Russian, and consequently are less likely to give unambiguous information about sentence structure.

There are, moreover, no relative pronouns in Japanese, and subjects and objects are often omitted when they are 'understood'. Thus, the literal translation of a sentence such as 'I saw the man whom we met yesterday' could be rendered in Japanese as 'yesterday met man saw', where 'I' and 'we' are understood and, lacking relative pronouns, the relative clause is placed in front of 'man'.

Of course, these linguistic idiosyncracies cause no more confusion to the native Japanese speaker than the many oddities of English cause problems to the native English speaker, but the fact that the fundamentals of Japanese and English are different in so many ways means that a much deeper understanding of language than a literal word replacement procedure is required for successful machine translation.

The upshot of these linguistic differences is that translation by computer requires a deep understanding of language and a serious consideration of *meaning* in conjunction with the rules of grammar of the languages being dealt with. Perhaps as a consequence of the great linguistic distance between English and Japanese, Professor Nagao was one of the early advocates of a semantics-based analysis of sentences. Current developments emphasize the (psychologically realistic) parallelism of simultaneous syntactic and semantic processing.

In this English translation of Nagao's original Japanese text, examples of translations between Japanese and English include (i) the original Japanese text, (ii) its transliteration into the Roman alphabet, (iii) a literal translation, and (iv) the final translation into proper English. Using the above example:

昨日会った叔父さんを見た。

KINO ATTA OJISAN WO MITA

(we) yesterday met man (I) saw

I saw the man whom we met yesterday.

The reason for this level of detail is to illustrate a point repeatedly made by Nagao and well-known to educated Japanese, but not obvious to native-speakers of English. That is, the translation between languages from unrelated linguistic families demands extensive rearrangement of words and phrases. Although the extent of changes in word-order appears at first consideration to be a disadvantage to the translator, in so far as a simple word-replacement procedure simply will not suffice, the impossibility of such simple techniques for translating into and out of the Japanese language has forced Japanese researchers to return to basic issues concerning the nature of translation. Instead of relying on the accumulation of a sufficient number of syntax-based rules for translation, they have faced up to the inevitability of dealing with semantic information.

Of course, semantics has not been ignored outside Japan, but the futility of a translation process based fundamentally on word-replacement is far more obvious in translating between Japanese and English than, for example, between French and English. In other words, the necessity of achieving a semantic understanding of the source text for translation is more apparent the greater are the syntactical differences between the languages. In this respect, the Japanese language, which is in an isolated linguistic family with only tenuous similarities with Korean and Chinese, has proved to be an advantageous starting point in considering the problems of machine translation. Once the general linguistic problems involved in the translation between Japanese and any of the Indo-European languages have been solved, it is thought that new problems should not be encountered in translating between other pairs of more closely related languages.

Following the realization that translation based solely on syntax will not be possible, recent efforts at bringing semantics into the translation process have not merely been the *post hoc* appending of some semantic information onto a syntax-based translation. Instead work has been done aimed at using semantics in parallel with what might be called 'traditional' syntactic techniques. From a psychological perspective, the parallelism between syntactic analysis and semantic analysis is of particular interest in light of the parallel processing of language in the cerebral hemispheres, where it is now known that, in addition to the phonological, syntactic, and denotative processing which takes place in the left hemisphere, the right hemisphere plays an important higher-level semantic role in placing the literal and denotative understanding of the left within a wider context.

It remains to be seen how deep the analogy between the human brain and computer systems for natural language processing may go, but it is certainly of interest that, in both realms, semantics is no longer considered to be the last of a dozen or so stages in language processing, but the second of two linked but separate processes. If the separate but simultaneous handling of these two aspects of language proves to be a crucial development in the evolution of machine translation systems, it may be possible to say that the difficulty of translating into and out of Japanese has actually been an advantage in coming to grips with fundamental issues in translation between any two natural languages.

An analogous situation in which the difficulties of the Japanese language have proven to be advantageous can be seen in the realm of dot-matrix printers. At the beginning of such print technology, it may well have been obvious to all that high-resolution dot-matrix printing would be desirable to produce legible print. From the Japanese perspective, however, a higher-resolution printing technology was not a luxury, but an absolute necessity. The seven-by-seven matrix which suffices to produce distinguishable letters of the roman alphabet was seen from the outset to be insufficient for producing legible *kanji*. The earliest 16-by-16 dot-matrix printers were thus produced in Japan, and a large percentage of the current generation of so-called near-letter-quality, 24-by-24 dot-matrix printers also come from there. The intractability of the complex ideograms thus became an important motivation in developing a printer technology to handle them.

It is not yet possible to say whether or not the idiosyncratic features of the Japanese language are such that Japanese efforts in machine translation will lead to significant advances in linguistic theory and then domination in the market-place for machine translation systems. It is nonetheless clear that there is more motivation for developing useful machine translation systems in Japan than in most parts of the world. Readers of this book will undoubtedly sense the breadth and depth of that motivation.

I would like to thank William J. Hutchins of the University of East Anglia and Karim Said of Oxford University for their comments and criticisms of the translated text.

Oxford   *Norman D. Cook*
1988

Contents

Preface to the English edition

Progress in research on machine translation systems in Japan

It is likely that objections would be voiced from many quarters to an assertion that Japanese research in machine translation is the most advanced in the world, but this is indeed the case. Careful reading of this book should convince most readers of this.

Due to the large differences between the structures of languages from completely different linguistic families, such as Japanese and English, and to related cultural differences, translation requires the use of completely different linguistic expressions in the two languages. The elucidation of translation techniques which take such factors into account is a major undertaking. As a consequence, the research in machine translation done in Japan over the last 30 years has encountered a huge number of obstacles. Many of these have been overcome, leading to the development of today's sophisticated systems. These problems have been of an order of magnitude greater than those encountered in translation among, for example, the various European languages. This is not to say that all the difficulties have been resolved, but the many techniques which have been devised to bridge the gap between Japanese and English will undoubtedly prove suitable for applications among the European languages or, indeed, between any two natural languages in the world.

Over the years, a large volume of books and technical literature from the West has been translated into Japanese. A great deal has thus been learned by the Japanese — leading to the modern state of Japan today. It is therefore partly a matter of obligation for Japan to return the favour by having many of its written works translated into the various Western languages. This reason alone makes machine translation essential.

It was in this spirit that the Science Council of Japan set out to develop a machine translation system (code-named the Mu-System) which will mechanically translate abstracts from the

Japanese scientific literature into English, and to present that information to the world as part of a large scientific database. For the four-year period from 1982, we took charge of that project and built two machine translation systems which could translate scientific abstracts from Japanese into English and from English into Japanese. We thus demonstrated the practicality of such research, and as a direct result of that work the Mu-System is scheduled to be added to the database search system run by JICST (Japan Information Center of Science and Technology). This will create a system which, sometime during 1988–89, will make it possible to inquire in English from anywhere in the world and receive an English-language abstract of any specific piece of Japanese scientific or technical research.

Commercial machine translation systems in Japan are also being vigorously developed. Businesses normally operate with the world market in mind, and every time a new product is introduced or an old product is updated, a huge volume of technical material related to its use and maintenance must be translated into various foreign languages. For commercial reasons, such translation must be completed at virtually the same time as the original Japanese documents. Translations must therefore be done in a short period of time, and a dependence upon machine translation appears to be an eventual certainty. Currently all such technical translations are done by human translators, and it has been estimated that sums in excess of 100 billion yen (400 million pounds) per year are spent on this alone. If even half of that sum can be saved by machine translation, a considerable saving will have been made.

It is anticipated, moreover, that the volume of such material requiring translation will continue to increase rapidly, and other uses of machine translation, such as the translation of letters and in the teaching of English, are also likely. Beyond the confines of machine translation proper, the need of the information society of the near future for computers to handle language and for free interaction in natural languages between man and machine, will require considerable development of natural language processing techniques.

For this reason alone, many of the Japanese businesses involved in information technology, particularly the computer manufacturers, are enthusiastically researching machine translation systems. Even if they do not realize a profit on early commercial systems of this kind, they understand that it is a useful exercise to accumulate expertise in techniques needed in the era of information. There are some ten companies in Japan which have already marketed machine translation systems, and there are at least ten more preparing to launch such systems. It is therefore extremely likely that, in a mere two or three years, a large number will be available commercially. In the market-place, of course, there is fierce competition, which will continue to stimulate the development of better systems and lead to rapid progress in the quality of machine translation as a whole. Currently available systems deal almost exclusively with English–Japanese and Japanese–English translations, but work is being done on Japanese–Korean, Japanese–German, and other translation systems. Future developments will no doubt include other languages as well.

Amidst such activity, the Japanese Government has kept an eye on developments in machine translation and has actively encouraged research on the various languages of South-east Asia, including those of China, Thailand, Malaysia, and Indonesia. Not only would the realization of such machine translation systems facilitate the transfer of various technologies to South-east Asia, but they would also be of value as manifestations of goodwill among these countries. The Japanese Government is consequently waiting to see how machine translation develops over the next few years, before full-scale projects are initiated.

It is well known that dictionaries in a machine translation system are extremely important and that the cost of constructing large, high-quality electronic dictionaries is considerable. It is therefore essential to build a so-called neutral dictionary which can be used by virtually any machine translation system. The Japan Electronic Dictionary Research Institute was established for collaborative efforts in the construction of such dictionaries. The joint efforts of the National government and private enterprise have harnessed the talents of many researchers from related fields. Research into an interpreting telephone system has also been initiated, to enable real-time voice recognition and machine translation (interpreting) of conversational phrases — ultimately aimed at inclusion in the international telephone system. This is discussed further below. Clearly, such projects will require world-wide collaboration on a large scale.

The place of techniques for natural language processing in future information industries

Computers are a fusion with and unification of communications technology at both the hardware and software levels, and computer systems will undoubtedly enter every corner of future society. When that day arrives, the most important technology will be specifically concerned with neither hardware nor software, but with what I have been advocating for many years: 'information-ware'. In other words, the central problem will be how the informational signals sent by human beings will be mechanically processed, transmitted, stored, and then recalled in a form which can be interpreted by other human beings. The essence of infor-mationware is therefore how information can be efficiently stored in a computer and activated in response to the various demands of its users.

There is in fact a wide variety of forms for information to take, including writing, speech, and visual images, but objectively, the most accurate means for transmitting and receiving information is writing. For this reason, of the various aspects of information-ware, linguistic information and the techniques for processing it will be the primary technology at the heart of the information society. Such technology might be called 'language engineering', and the industry which it will spawn will be 'language industry'. It can be imagined that the 'language industries' of the 21st century will become the largest industries of the future and will have a much greater scope than they have today.

Many Japanese businesses have already come to realize the importance of such developments and are therefore making considerable efforts to acquire this technology. Among the actual devices which fall within the realm of language engineering and language technology are machines for recognizing handwritten language, machines for recognizing speech, interactive systems between man and machine, and sophisticated man–machine interfacing techniques. At the heart of all such devices are the techniques which are at the core of machine translation. More precisely, it can be said that machine translation is the synthesis of many facets of natural language processing into one coherent system — the input and output of words, morphemic and syntactic analysis, semantic and contextual processing, and the pragmatics of conversation between the speaker and listener. It has recently become possible to evaluate on fairly objective grounds the quality of the translations produced by such systems from a wide variety of source documents. The possibility of unambiguous identification of the successes and failures of such systems has created the further possibility of making verifiable improvements in them. Thus, for the acquisition of the techniques of natural language processing, the best possible means is the study of machine translation systems.

Broadly speaking, there are two approaches which can be taken to developing a new technology. The first is to begin with the simplest problems and gradually proceed to the solution of more difficult ones. The second is to tackle the difficult problems from the outset, the solution of which will mean that the easier problems are also solved. The techniques of character recognition devices are an example of the former approach, and those of natural language processing an example of the latter. In the case of visual character recognition, developments have taken place over a period of already more than 20 years — starting with numerals and proceeding to the roman alphabet, the Japanese *kana* alphabets, the Chinese ideograms, and finally to handwritten ideograms. As a result, it is now possible to read by machine a Japanese-language text containing over 3000 ideograms.

Once the extremely difficult problems in machine translation have been solved, the various techniques thus developed can be applied to other fields in natural language processing. Efforts are now being made to develop the highest-level techniques by tackling one of the hardest of translation tasks — that between Japanese and English.

An analogous situation can be found in the development of the railway system in Japan. Until recently, many difficulties were encountered as a result of the introduction of a narrow-gauge railway system in the Meiji Era, and various technologies in railway engineering have been perfected since then. The diversity of those efforts led to Japan being the first to succeed in developing a high-speed 'bullet' train. In machine translation, we are

faced with a difficult problem in the automatic translation between Japanese and English, with their completely different
cultural settings and linguistic constructions, but in the process of
overcoming the unusually difficult problems encountered in this
kind of translation, a very sophisticated technology is evolving.

Machine translation systems: the choice between pivot language and transfer techniques

The process of machine translation can be divided into three
distinct stages: (i) the analysis of the source language text; (ii) the
transfer of the expression of the results of the analysis into an
expression related to the target language; and (iii) the generation
of the text in the target language. The so-called pivot-language
method is one in which stage (ii) is not required, but rather
translation proceeds directly from the results of stage (i) to generation, stage (iii). When stage (ii) is employed, it is referred to
as translation using the transfer method.

The recent emergence of questions concerning the construction
of machine translation systems which can translate among a variety of languages has led to debate about which of the two
methods is better. It has been emphasized by advocates of the
pivot-language method that when translating among 'n' different
languages, the transfer method requires the construction of n ×
(n−1) transfer routines, whereas such modules are totally unnecessary using the pivot-language method. For this reason, pivot
language-based machine translation systems have the advantage
of being a good deal simpler.

What about the quality of the translation? In the routines of
stages (i) and (iii), only information concerning, respectively, the
source and the target languages is required, and there is no need
to consider the properties and idiosyncracies of the other (target
or source) languages involved in the translation. Therefore, when
the pivot-language method is used, the results of the analytic
stage must be in a form which can be utilized by all of the
different languages into which translation is to take place.

For example, although they are infrequently used in European
languages, in Japanese there are many words of respect and
politeness which reflect the social positions of the speakers, as
well as distinctly male or female expressions which lie at the heart

of Japanese culture. These are factors which must be considered when translating between Japanese and European languages, but they are far less important when translating solely within a European context.

It is important to note that in the case of translation through a pivot language, since it is necessary to decide upon expressions in the pivot language which will be appropriate for translations among all possible language combinations, a model must be used in which *all* of the factors found in *all* of the languages are included — including linguistic expressions dependant upon sex, social status, degree of respect, and so on. Even when those factors are not explicitly expressed in the target language, they should be inferrable from the context, from the psychological state of the speaker, or from the cultural background of the language. The pivot language must be extremely well constructed if it is to include such subtle factors. Since we do not in fact know into which language the results of the analysis of the source document will eventually be translated, it is essential to produce expressions in the pivot language which contain results which are as fine grained in every respect as the target language.

This level of subtlety is a practical impossibility, and with regard to translation between European languages, it is also unnecessary. Clearly, the degree of subtlety needed in the analysis of the source text depends on the target language. For the translation of straightforward official documents among the languages of Europe, it would conceivably be possible to use the internal structure of a language like Esperanto as the pivot language. However, since an Esperanto-like language does not deal with the expressions of respect and politeness found in Japanese, using such a language as the pivot language would limit the linguistic expressions which the system could handle, and consequently greatly reduce the range of the natural languages which the machine translation system could deal with. In essence, at the current level of understanding of linguistics and natural-language processing techniques, it is impossible to find a pivot language which can be used for translations among all languages. Moreover, analysis of source texts to the level of complexity described above remains an impossibility. As a consequence, machine translation using the pivot-language method is suitable for translations in which only the basic meanings of the source text are to be reconstructed in the target language. When translation of subtle nuances is also to be attempted, the transfer technique must be used.

Another point of debate regarding the model to be used in machine translation is whether the system should rely fundamentally on syntax or semantics. Early machine translation systems relied solely on syntax, and all structural analysis was done on the basis of rules of syntax. What was learned from such systems was that, even in the case of extremely simple sentences, a large number of logically possible results of analysis were produced, and choosing among the possibilities on syntactic grounds alone could not be done. This is the problem of so-called syntactic ambiguity. A related problem concerns languages, such as Japanese, in which the word-order can be rearranged rather arbitrarily. Although the meanings of the various permutations of a sentence will be essentially the same, an extremely large number of grammatical rules are needed to cover all possible word-orders. The more that such detailed grammatical rules are included in the system, however, the greater is the variety of grammatical constructions covered and, as a consequence, the greater is the danger that the fundamental function of such rules will lose their practical value.

Clearly, these problems arise because semantics is not brought into consideration. For example, although we may frequently say 'pretty girl', we cannot often use the phrase 'pretty boy' — not because of grammatical considerations, but because of semantic implications. On the basis of such ideas, it has become recognized that sentence analysis must be undertaken at least partially on the basis of semantics. One form of grammar in which the grammatical rules also handle semantics is the so-called case grammar. Instead of being based upon the word-order in a sentence, case grammar focuses on the role of nouns in relation to verbs. By means of the introduction of this concept of resolving questions of syntax on the basis of the meaning relations between nouns and verbs, it has become possible to devise machine translation systems in which semantics takes a central place. That is, by joining case grammar techniques with the pivot-language method discussed above, the structural analysis of sentences results in a semantic representation that can then be translated into expressions in the pivot language. By definition, the pivot language, in turn, has a structure in common with all natural languages. Analysis and translation on the basis of such semantic considerations has been found to be unquestionably more advanced than analysis and translation on the basis of syntax alone. Great emphasis has been put on the fact that many natural languages can then be handled by means of a single, unified pivot-language method.

However, this solution contains many problems. First, a theory of semantics of sufficient detail to enable unambiguous handling of meaning has not yet been devised, and it cannot be said that debates concerning meaning have a firm foundation in formal logic. On the contrary, hypotheses about meaning are as numerous as the number of researchers, and it can be argued that research in this field has just begun.

The theory of meaning devised by Montague is extremely attractive, and some believe that machine translation will be possible on that basis alone. Montague's theory, however, is based fundamentally on meaning in the sense of the theory of symbolic logic, and considers the expression of a linguistic statement as a process corresponding to the truth or falsehood of events in the actual world. In contrast, the meanings which human beings normally deal with have much broader implications, and it is this latter sense in which meaning must be handled in machine translation systems. It is still unclear in what configuration such broad meanings can be stored and used in a mechanical system. Furthermore, there is little prospect that sentence analysis by means of so-called categorial grammar (which uses Montague grammar) will be able to deal with the complex sentences which need to be handled in machine translation. For this reason, it is probably fair to say that research on machine translation using Montague grammar has ended in failure.

In contrast, when case grammar is used proficiently, dictionary information must be provided on the meaning of each word and on how it can be used in relation to specific verbs. For this purpose, a large number of fundamental semantic units, the so-called semantic primitives, must be prepared and used to express the meaning of every word in the dictionary. Needless to say, such dictionary construction work requires a great deal of labour, and it is often difficult to represent accurately the meaning of a word solely by means of its semantic markers. In other words, this technique leads to a great deal of inaccuracy.

Moreover, in the pivot-language method based on semantics,

the information contained in the word-order of the source text is usually not maintained in the pivot-language representation and is lost in translation into the target language. Word-order, however, can have semantic implications. For example, a phrase which the speaker wants to emphasize or the scope of the theme of the sentence can be expressed implicitly through word-order, and word-order is deeply involved in problems related to the ease of reading or understanding a text. Consequently, it has been found that if translation is done ignoring syntax and relying solely on meaning, sentences which are extremely difficult to understand will often be produced. Since it is not possible at the current technical level of linguistic analysis to extract this kind of implicit information accurately and unambiguously, it is desirable to maintain the word-order and style of the source text as far as possible, and to make only the unavoidable syntactic changes — thereby maintaining this type of information in the sentences of the target language. As a matter of fact, it has been reported that sentences translated by machine solely on the basis of syntactic analysis are notably easier to read and understand than translations using a pivot-language method based on semantics.

In the light of findings such as these, the preferred language-processing model is one in which syntax and semantics are handled in parallel. Since semantics alone does not normally provide trustworthy results, a method is used in which the results of syntactic analysis are emphasized. Only for problems of disambiguation and analysis of sentences in which subject or object have been omitted (as is often the case in Japanese), should case grammar or semantics be relied upon, and even then only in a fundamentally supplementary role.

The essence of language is normally to express new ideas in new forms. This means linking words which had not previously been linked in this manner. The interpretation and understanding of the new expression by the human listener depends on (i) the meanings of the words and (ii) the common-sense knowledge held by the listener. It should be noted, however, that understanding depends crucially on (iii) this new combination of words, that is, on the basic syntax of the sentence. A sentence such as 'The stone is flying in the soup' is not permissible in normal semantic analysis, and indeed semantic machine translation of such a sentence would lead to complete failure. A human being, on the other hand, will be able to handle such a sentence using his imagination. In such a case, the syntax of the sentence will be of extreme importance in arriving at a suitable interpretation. Guesses will then be made about the semantic relations, based upon syntactic relations, and even if it is an impossible situation, the imagination will construct a plausible interpretation. Such extremely flexible functions are an inherent part of how the human brain processes natural languages.

These considerations strongly suggest that, in the analysis, transformation, and generation of sentences, syntactic rules should be given a primary role. It is also extremely important that the syntactic and semantic analyses do not occur in isolation from one another, but that they should occur almost simultaneously. In this way, a large volume of the results of syntactic analysis will not be produced, but instead the meaning of the sentence will be arrived at rapidly. This approach is particularly important in analyses of a language such as Japanese, in which word-order is relatively unconstrained.

In a typical engineering system, the scope of the application would normally be clearly defined at the design stage. For example, in the design of an automobile, a vast number of factors are predetermined, such as the distance the car can travel from the time of applying the brakes, the durability of the car on roads of a given smoothness, and so on. Within these set parameters, the highest-quality vehicle is then produced. This concept of defining the parameters of the system is one of the fundamental ideas behind all engineering, but the circumstances surrounding the design of a machine translation system are unfortunately not conducive to the determination of all parameters prior to production. Futhermore, it is simply impossible to provide definitions of the scope of linguistic phenomena, to define clearly what language is and what constitutes legitimate and illegitimate linguistic expressions.

Consequently, it must be possible to add new analytic components to the system to allow for the handling of expressions which had not been considered at the time of the original design of the grammatical rules of the system. There is therefore the possibility that the entire system will require rebuilding in the light of the model which the new phenomena suggest. For this reason, the framework for a machine translation system must be extremely flexible and robust, allowing not only for the handling of a wide variety of phenomena, but also allowing further additions and modifications to the system. The system itself cannot therefore be a closed one — typical of most engineering projects — but must remain open, such that new functions can be appended as required. A system of such flexibility will require considerable ingenuity in developments at the software level.

Machine translation systems are necessarily huge. In order for the system to be coherent and capable of self-correcting evolution, it is desirable that the entire system be built around a single, unambiguous (open-system) theoretical framework. At the same time, however, language is a product of unbridled human mental activity and is in a state of continual change. Linguistic activity does not take place within a single mathematical system or a single logical framework. Therefore, even if a clear and unambiguous language-handling theoretical framework could be devised, it can be expected that exceptional and unexpected linguistic phenomena would appear. There is no doubt that, for example, lexical functional grammar and generalized phrase structure grammar are excellent grammars for explaining and handling the fundamental structures of language, but they are most certainly too simple to explain all of the phenomena of a living language, which people manipulate freely. Moreover, it must not be forgotten that machine translation is not concerned with one language, but with transformations among many languages, which, as mentioned above, entail many syntactic and semantic changes.

Every word in a given language has a specific meaning and contains unique information which makes its replacement by other words often impossible. Because there exists such a huge volume of linguistic phenomena associated with every individual word, each must be stored in the dictionary for machine translation, together with the rules for its use, rather than relying on general theoretical principles. Means by which this information can be efficiently used must consequently be devised.

Stated bluntly, language is a massive conglomeration of exceptions, and any competent machine translation system must therefore be able to incorporate the exceptional aspects of language simply and easily at every stage in its construction and function. In other words, the entire structure of the system must be open ended and evolutionary. The Mu-System at Kyoto University was designed with this principle firmly in mind. During its construction and use, when a new or exceptional phenomenon is encountered, it is possible to insert that phenomenon directly into the system. The time required for improving the quality of such a system in this piecemeal fashion is of course extremely lengthy, and we anticipate that any rudimentary system will require years of refinement before it is truly useful. Such machine translation systems are virtually never fully completed.

Future developments in machine translation systems

There are still people who foolishly claim that machine translation systems are in principle no further advanced than those available at the time of the *ALPAC Report* in 1965 (see Chapter 1). Such a claim is fundamentally the same as (and as wrong as) saying that the basic principles of computers have not evolved since the days when von Neumann first programmed his calculating machine. Computers have of course been developed to an astounding degree in terms of hardware, software, and applications since then, and similarly the developments in machine translation have been remarkable. These have included developments in the fundamental ideas concerning software systems, grammars which incorporate a considerable amount of semantics, translation by means of compex structural transformations, the efficient use of a mammoth dictionary of words in common use, the storage and retrieval of extremely detailed linguistic information for use in translation, and so on. Many of these developments were simply unanticipated in the early 1960s.

Moreover, the market for machine translation has grown by an order of magnitude since the 1960s, and the demand continues to grow. Users are learning how to exploit machine translation systems to their fullest. In several years' time, it will therefore be possible to build systems which have been adapted to the needs of specific users. If still lacking in elegance, such systems will be fully practical tools for translation. Developments in the coming years can be expected to lead to the rapid infiltration of machine translation systems into society at large. Let it be emphasized that at the heart of such a forecast is the fact that, in Japan, a large number of businesses are involved in the development of commercial machine translation systems.

If, however, we ask whether the current techniques behind machine translation are sufficient or complete, the answer is more pessimistic. In order to determine the referents of pronouns and abbreviations, techniques have been devised for current systems to backtrack through a certain number of sentences for relevant clues, but as a rule translation is undertaken one sentence at a time. Such systems manifestly do not embrace the methods used by human beings, who normally read through and understand the meaning of several sentences before undertaking the actual translation process. Moreover, it remains impossible for machines to arrive at appropriate translations in the light of inferences drawn about the intentions of the speaker or underlying cultural factors. These issues have only recently become the focus of research in linguistics and cognitive science, and a long period of intense study will certainly be required before answers are arrived at which are concrete enough for implementation in engineering systems. Nevertheless, once the aims of such research have been clearly defined, there is every reason to expect that significant results can be achieved. This is in fact the general area within cognitive science where we can expect the most interesting new developments over the coming years.

In the light of such considerations, a Machine Translation Summit was held in Japan in September 1987. Those responsible for organizing research on machine translation in Europe, America, various Asian countries, and Japan gathered with the researchers and users of such systems and debated the direction of research, the needs of end-users, and the current trends in commercial machine translation systems in an international context. Some consensus was reached regarding the expansion of machine translation in the coming years. I presented my own conception of the past and future stages of machine translation (Fig. 0.1).

In March 1986, the International Institute for Advanced Telecommunications Research was established, and research was begun into the cognitive issues of human communication, together with automatic telephone translation (interpreting). The latter topic is concerned specifically with means by which one can speak in either Japanese or English into a telephone, and have the words translated and spoken to the listener in the other language. The aim of the Institute is to undertake the basic research in speech recognition, machine translation, and speech synthesis

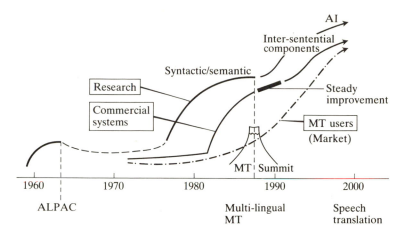

Fig. 0.1   A summary of the history of machine translation.

required to enable the construction of a working prototype within 15 years. It will be necessary to consider such communication on a basis quite different from the translation of a correctly constructed written text, since telephone conversation entails many additional complications. These include the facts that conversation is normally fragmentary, grammatically incorrect sentences are often produced, various insertions and interruptions are common, and understanding is frequently reached in the middle of a sentence. Since (i) conversations often take the form of questions and answers, (ii) conversations are fundamentally interactive and do not necessarily entail the logical development of a single theme, as does a good written text, (iii) conversations often consist of incomplete and fragmentary sentences, and (iv) even if one cannot fully interpret the meaning of a statement, one may be able to give a suitable answer if one understands the basic outline of the conversation, it seems appropriate that the analysis of converstional sentences should be performed with emphasis on the semantic rather than the syntactic perspective. Therefore, it may be best to employ a semantic network based upon case grammar for the internal structure of the analytic results, thus also allowing for translation among a variety of languages. One additional problem faced by researchers in automated telephone interpreting is of course voice recognition. If the realm of conversation is restricted to appropriate limits and sufficient general knowledge pertaining to that realm is supplied to the system, the difficulties posed by such a system should not be insurmountable.

In addition to such adventurous research, in April 1986 the Japan Electronic Dictionary Institute was established and work begun on the construction of a neutral Japanese–English dictionary — focusing on concepts — for use in a broad spectrum of machine translation systems. Already a private company has embarked on the construction of a machine translation system capable of translating between Japanese and Korean. Collaborative research projects with the respective countries have also initiated the development of machine translation systems which can handle the Japanese, Chinese, Thai, Malayan, and Indonesian languages. The future of machine translation systems is truly exciting.

Dedication

At the conclusion of writing the Japanese text of this book in the autumn of 1985, I was saddened to learn of the sudden death at the age of 56 of Professor Bernard Vauquois of Grenoble University, for many years one of the world's leading figures in machine translation. He had long been the Chairman of the Association for Computational Linguistics and had kindly invited many Japanese researchers to Grenoble for collaborative work. For me, the death of this academic rival — in the best sense of the word — is a true loss. I would therefore like to dedicate this book to his memory.

1 The dawn of machine translation

The Sputnik shock

I have before me a volume of research reports on machine translation,[1] given to me by Hiroshi Wada, one of the early leaders in machine translation in Japan. Wada was the Chief of the Electrical Testing Center (now known as the Electrotechnical Laboratory) of the Industrial Institute of the Ministry of International Trade and Industry (MITI), and was one of the earliest developers of Japanese computers and computer applications. He was also one of the original workers in research into machine translation, starting in the latter half of the 1950s.

The research volume which he gave to me is entitled *Research on Mechanical Translation* and had been submitted to the Leader of the US House of Representatives by a Special Committee on Mechanical Translation within the Committee on Science and Astronautics of the US Congress. It contains the material discussed on 28 June 1960 in the 86th Congressional Meeting, and covers the history of American research on machine translation from the time Warren Weaver declared the feasibility of machine translation in 1949, until 1960. It also lists sources of research monies, details the current state of American research as of 1960, and contains a thorough analysis of developments to that time in machine translation throughout the world, including Japan. Finally, there were fully 13 pages of conclusions and recommendtions to Congress.

My own entry into the world of machine translation was immediately after the completion of my graduate work in 1963. I was in fact unaware of the existence of that Congressional Report until Dr Wada showed it to me, but several books specifically on machine translation had already appeared. I learned a great deal from my reading about the history of machine

translation through the 1960s, and since then I have been involved in research in this field. Rereading that Congressional
Report of 1960 today, I have many strong feelings, the reasons
for which will become apparent. Briefly, however, it is because
the road towards machine translation has been filled with so
many dramatic developments, great hopes and great failures, eras
of hard work and eras of great expectations.

The news on 4 October 1957, that the Soviets had successfully
launched the world's first artificial satellite, the Sputnik, came as
a tremendous shock to the Japanese people, but it was a far
greater shock to the Americans. Believing that America had
fallen behind the Soviet Union in technological capabilities, the
US Congress immediately passed the National Defence Education Act of 1958 in order to revamp scientific education and to
encourage technological development throughout the United
States.

As a direct result of the sustained efforts that followed the act,
the Americans were the first to land a manned spacecraft, the
Apollo, on the surface of the moon, in 1969. The strength of
American technology was thus exhibited to the entire world. In
addition to recommendations on education and research, the
National Defence Education Act of 1958 had proposed a technical scientific information service. More importantly, there was a
statement to the effect that for American scientific organizations
to be able to make use of technical and scientific information
from throughout the world, plans should be drawn up for the
development of new and improved methods for such purposes,
including 'mechanized systems'. In this manner, research on
machine translation in America gained official recognition in
Congress and the Federal Government in general. Related plans
were not confined to the scientific organizations of America, but
were also actively pursued by the CIA, the US Military, and the
National Bureau of Standards. Research into machine translation
was thus started at tens of locations in the USA, including universities, research institutes, and private industry. It can therefore
be assumed that the subsequent recommendations and reports on
machine translation given to the 86th Congress in 1960 were
made in an atmosphere reflecting the optimistic developments in
research at many locations.

From this report, one does not get a sense of the rose-coloured dream which more fervent researchers in machine translation had in the early 1950s. However, the clearly superlative fundamental research accomplished over those first 10 years can be summarized, by saying that there were suggestions that a genuinely practical machine translation system could be built, and that in order to build reliable machine translation systems, a National Machine Translation Institute would be required. Indeed, it is clear that in 1960 no one could foresee the decisive negative declaration on this subject which was to come some five years later.

The very beginning of research on machine translation can be traced to an idea proposed by Warren Weaver of the Rockefeller Foundation in 1946, only one or two years after the construction of the first computer. He suggested that if the code-deciphering techniques developed during the Second World War were used, computers would be able to recognize the fundamental aspects of all known languages.

Andrew D. Booth, who was consulted on this issue, argued that 'dictionary translation' could be done, provided only that a large enough computer memory was available. Further developing these thoughts, Weaver produced a memo entitled *Translation* in July 1949, and distributed it to the leading figures in computing in America. It suggested the possibility of translating natural languages by means of computer. As a direct consequence, research on machine translation began in earnest at the University of Washington, UCLA, and MIT, and with the aid of funding from the Rockefeller Foundation, the first conference on machine translation was held in the spring of 1952. At that meeting it was suggested that computer systems capable of doing word-for-word translations would be possible in the near future, and that research on how sentence structure might be elucidated should begin immediately. In January 1954, the results of collaborative research between Georgetown University and IBM were revealed. It was shown that, in a machine translation experiment in Russian-to-English using a limited range of words, successful translation was achieved with a vocabulary of 250 words and only six rules of syntax!

With the attempts at machine translation at various universities and the appearance of a journal entitled *Machine Translation*,[2] the potential value of machine translation was gradually recognized both within academia and in society at large. The US Government provided universities and research institutes with research funds for related work, and by 1957 a considerable volume of research had been completed at many locations in the US. As a result of the 'Sputnik shock', it was thought essential to translate rapidly and in large quantities the technical and scientific information available in foreign languages, particularly Russian. In retrospect it is clear that this was a period of peak interest in machine translation research.

The years immediately before and after 1957 were a truly amazing era. Research which ultimately led to significant progress was being done in a number of fields. In psychology, Jerome Bruner was advocating ideas about the psychology of perception and George Miller was emphasizing that the short-term memory of human beings could hold only seven (plus or minus two) items at any one time, for example, seven unrelated words. In linguistics, Noam Chomsky was advocating a new linguistic theory which ultimately had a considerable impact. He had done research on a new style of linguistic theory under the guidance of Zellig Harris at the University of Pennsylvania, but he then moved to MIT, where he studied mathematics and fundamental computer theory, including the theory of automata and theory of computability. In 1957, he published his highly influential linguistic theory, called transformational generation grammar, which was quickly recognized to be suitable for implementation in computer systems. Since 1960 and continuing up to today, machine translation research based upon Chomsky's theory has been actively pursued.

Research on machine translation in the Soviet Union began in 1955 in response to the demonstration of Russian–English machine translation by Georgetown University and IBM. Research was begun at a large number of organizations, particularly within the Soviet Academy of Sciences, and by 1960 a much larger number of researchers were engaged in machine translation in Russia than in America. This remains the case today, but it is largely because the Soviets do more translating than anyone else. A huge volume of the world's literature, scientific and otherwise, is translated into Russian, and a well-organized system has emerged for collecting the specialist terms of the individual fields of research to facilitate translation work. However, the computer situation in the USSR has always been notably inferior, and there are few if any machine translation systems to which researchers have easy access. It appears, therefore, that most research in the Soviet Union has been of a theoretical nature.

Machine translation research also began in Europe in 1955. In England, the most significant efforts were made at the Cambridge Language Research Unit, where collaboration with US researchers was encouraged. In France, under the auspices of the Centre Nationale de la Recherche Scientifique (CNRS), an automatic translation research institute was set up at Grenoble University, where work has continued until today.

The beginning of machine translation in Japan

Research on machine translation was first undertaken at Kyushu University in 1955.[3] At the center of that work were the late Professor Toshihiko Kurihara and the current Professor, Tsuneo Tamachi. In 1955, Tamachi received a Science Research Grant from the Ministry of Education to build a dedicated translation computer, the so-called KT-1. This was an ambitious project for the production of a machine which would translate among the Japanese, English, and German languages. At that time, of course, computers capable of such work were not available, so they started the dual task of building a computer suitable for translation purposes and of learning how to do mechanical translations! By 1960 the system had been more-or-less completed.

Around 1957 research on machine translation was also initiated at the Electrotechnical Laboratory of the Industrial Institute of MITI, under the supervision of the electronics chief, Hiroshi Wada. A transistor-style computer was already under development at the Electrotechnical Laboratory, and Wada had the foresight to see how such a device could be used in the realm of language processing. They set about research on automatic character recognition and machine translation. That system was nicknamed 'Yamato' and was completed in February 1959. It was the first English-to-Japanese machine translation system available in Japan. An example of the rather rudimentary translation capabilities of this system is shown in Fig. 1.1.

I began my research on machine translation at Kyoto University in 1963, under Professor Toshiyuki Sakai, just after

This is the book which is mine.

コレ ガ ホン（ソレ ガ ワレ ノ モノダ）ダ。

KORE GA HON (SORE GA WARE NO MONO DA) DA.

this book (that I of thing is) is.

The computer does not forget whatever he learned.

computer ガ（カレ ガ マナビタ モノ ハ ナンデモ）ヲ ワスレナイ。

computer GA (KARE GA MANABITA MONO WA NANDEMO) WO WASURENAI.

computer (he learned thing whatever) does not forget.

Fig. 1.1 Examples of the translations performed by the first Japanese machine translation system, 'Yamato'. The *kanji* could not be handled by computer at that time, so English was translated into *katakana*, one of the Japanese phonetic systems [from Wada, H. (May 1982). Machine translation. *Information Processing*, 144–53. (Jpn.)]

completing my master's degree. I was then most interested in a question widely debated in the latter half of the 1950s, whether or not the computer would prove to have unlimited capabilities. An Englishman, Alan Turing, had already provided fundamental arguments concerning what computing is and what types of work computers can theoretically do, and he had already published his well-known model of computing, the so-called Turing machine. He is also credited with the theory of recursive functions within the framework of fundamental mathematical theory, and is largely responsible for outlining the theoretical possibilities and limitations of computing. At that time, however, little was known about how the mechanisms of information processing and decision-making in the human brain might be elucidated, or how computers could be made to have similar capabilities. For my part, I had a driving interest in questions about the degree to which the activity of the human brain could be achieved by computers. I then undertook research on two related problems:

(i) pattern recognition (specifically, how written characters could be read by computers in a manner similar to human reading skills), and (ii) machine translation (how computers could translate text in a manner similar to the human capacity for understanding and translating words).

As can be seen from this brief history, research on machine translation was taken up in many countries throughout the world in the mid 1950s, and in fact, many small machine translation systems came into being. Exaggerated claims and overoptimistic expectations were then widely disseminated, and the belief that translating computers would soon have the capabilities of human beings was common. Within this optimistic atmosphere, considerable research funding, was thrown at the problem of machine translation for several years, particularly in America.

Difficulties in understanding sentences

Unfortunately, machine translation did not turn out to be so straightforward that it could be solved simply by throwing money at it. As had already been discussed in the report presented to the US House of Representatives in 1960, translation involves many of the most difficult questions concerning the nature of language understanding and generation. It became increasingly clear that, in order to achieve translation by computer, a long period of fundamental research into unknown areas was required.

One of the early workers in this field was Susumu Kuno, who had studied at Harvard University after graduating from the Department of Linguistics of the University of Tokyo. Kuno applied the techniques of syntax analysis developed by Ida Rhodes, then of the National Bureau of Standards in Washington, DC. Rhodes's technique, called 'predictive syntactic analysis', had been used for analysis of Russian-language texts, but Kuno used it for the analysis of English. It does not take semantics into account, but relies solely on syntactic rules. Already by 1962 he had demonstrated that the syntax of a single sentence could give several different interpretations. For example, he showed that a simple sentence such as:

Time flies like an arrow

could be interpreted in several ways. 'Flies' is the third person singular form of the verb 'to fly', but it is also the plural form of the noun for the insect called 'a fly'. In addition to being a conjunction, 'like' is also a verb meaning 'to love' or 'to be fond of', and has the appropriate third-person plural verb form which would correspond to a plural subject (flies). Therefore, if there were a special kind of fly called a 'time fly', this sentence could be interpreted as meaning that these hypothetical 'time flies' are fond of a particular arrow!

Since many nouns in English can also act as verbs, this type of ambiguity does arise, causing confusion for the native speaker only infrequently, but potentially confounding an automatic machine translation system. In fact, Kuno demonstrated that a moderately sized sentence containing some 30 words can have more than 100 different results when subjected to syntactic analysis. He also pointed out that even in the nuclear test ban treaty between the USA and the USSR, similar ambiguities were present and would allow for different possible interpretations of the treaty itself.

These findings indicated in concrete terms that analysis of sentences solely on the basis of syntax was not good enough, and moreover, that the meanings which words have — that is, the realm of 'semantics' — must also be considered. Just at this time, I was wondering how the meanings of words should be incorporated within automatic translation systems, and Victor Yngve of MIT was doing research on Miller's hypothesis that the human brain can remember only seven (plus or minus two) items at any given time. In order to test this hypothesis in the realm of language, Yngve succeeded in implementing Chomsky's phrase structure grammar within a computer program and was experimenting with the production of sentences by computer. As a result of consideration of these experiments, I had become convinced that semantics plays a central role in the generation of structurally appropriate sentences.

From that modest starting point, I proposed in 1964 a method for investigating the usefulness of 'meaning tables' or 'semantic networks', which have been widely employed ever since. That is, I showed that by employing extensively interconnected word tables, the semantic relationships among the nouns in a sentence could be determined, particularly those relating to the main verb of the sentence. Only those sentences which were found to have appropriate meanings would then be generated, and the many grammatically sound but meaningless possibilities would be discarded. Since meaning tables are systems for classifying groups of words, it was necessary to set up a large number of appropriate semantic categories, such as: human beings, animals, plants, minerals, abstract concepts, and so on. This approach to language has since been combined with ideas concerning case grammar and today is used in many machine translation systems. This is discussed further below.

It became clear to me that for the analysis and translation of languages by means of computer, not only must the syntax which determines the structure of the sentence be considered, but also the complex relationships of meaning (semantics) must be dealt with. The study of linguistics, however, had at that time hardly begun to deal with the problem of meaning. An obstacle in the path of further research had therefore been reached, and there were few clues available on how to get round it.

The *ALPAC Report*

Amidst the world-wide expectation that a sophisticated human task such as translation could be performed by computers, research funding for such work was provided in America over a considerable number of years. Unfortunately, the emergence of practical machine translation systems did not follow. Perhaps as a result of this, Yehoshua Bar-Hillel, who had been active in the 1950s at MIT, where much of the research on machine translation in the US had been undertaken, and later at Jerusalem University, wrote a paper in 1960 declaring that fully automatic, high-quality machine translation was categorically impossible. With that pronouncement he quit research in this field. Together with various pessimistic results, such as those of Kuno demonstrating the difficulties in sentence analysis, it was inevitable that doubts would eventually emerge about the possibility of actually achieving machine translation.

Thereafter, the US Academy of Sciences set up an Advisory Committee on Automatic Language Processing. That was in April 1964. The Committee took more than a year to complete a survey of the current state and future prospects of machine translation research in America, and made a detailed study of societal needs for translation. They also undertook a comparison of the quality and costs of translations done manually and those done by machine. Their conclusions were that, at that time and for the near future, machine translation was and would remain notably inferior to human translation, in terms of both quality and cost. Those conclusions were packaged in the notorious *ALPAC* (Automatic Language Processing Advisory Committee) *Report*, which was then submitted to the National Academy of Science on 20 August 1965.

The *ALPAC Report* analysed the current state of translations, particularly technical-scientific translations, in the USA from many different angles. It discussed recent developments in machine translation in the USA and Europe, and made comparisons with human translations done in West Germany and Luxembourg using a so-called electronic dictionary. Its conclusions were that, rather than pursue machine translation itself, three lines of research should be followed. The first was to study ways in which manual translations could be done more quickly and economically; the second was to promote research on computational linguistics; and the third was to adopt the position that linguistics is a scientific discipline and to recommend fundamental research on language *per se*, quite aside from the immediate practicality of such research within the realm of machine translation.

The basis for these conclusions was provided in an appendix to the report, containing fully 20 subsections. Altogether the report contained only 124 pages, but it remains an extremely relevant document even today (the Table of Contents from the *ALPAC Report* is shown in Fig. 1.2). Suffice it to say that it has been necessary in all subsequent efforts in the realm of machine translation to start where the *ALPAC Report* left off.

As a direct result of the *ALPAC Report*, research funding for machine translation, which had reached 20 million dollars from the US Government alone in the preceding ten years, was reduced to virtually nothing. The majority of researchers were removed from their positions, and as a consequence the knowledge and experience of machine translation which had been accumulated in America over ten or more years was simply lost. From the perspective of today, it is clear that these losses were great indeed.

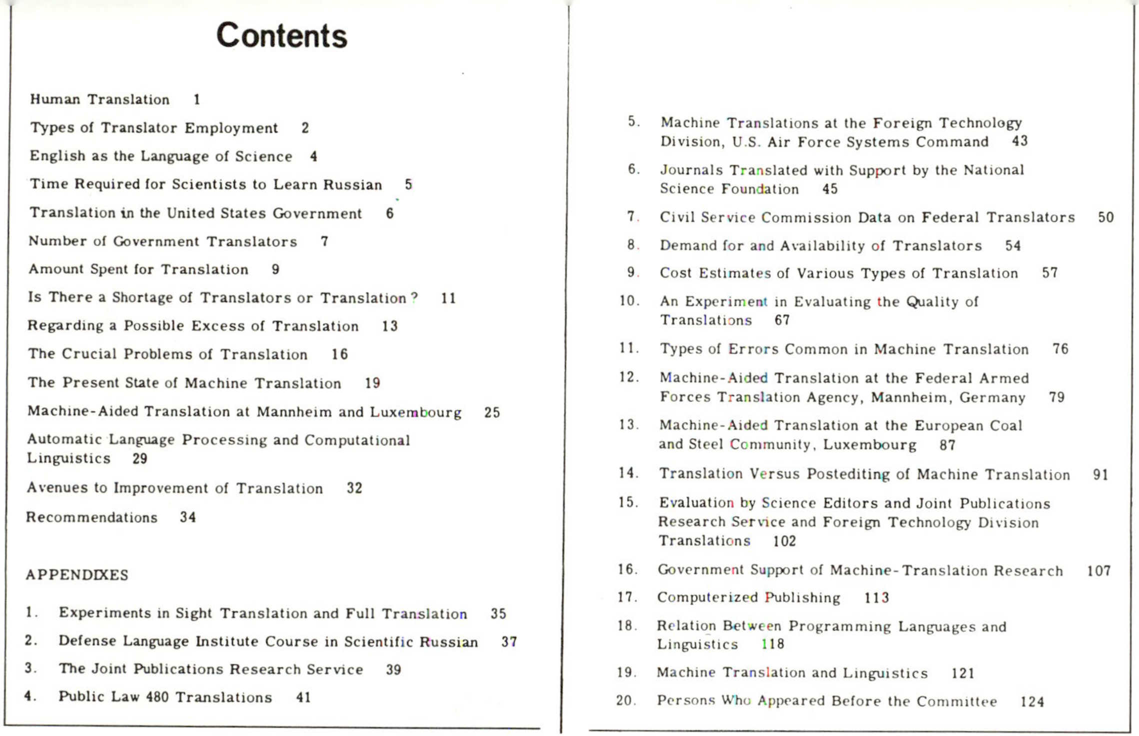

Contents

Fig. 1.2 The table of contents of the infamous *ALPAC Report*. The scope of the survey of machine translation undertaken by the committee is evident from the list of topics covered.

A period of reflection

If we consider the period from 1950 to 1965 as the first era of machine translation research, the ten-year period from 1965 to 1975 (that is, prior to the current era of activity) can be regarded as the Dark Ages of machine translation. Following the publication of the *ALPAC Report* in 1965, it became virtually impossible to do research in this field in America, simply because research funds were not available. Inevitably, this situation in the USA had repercussions abroad, and the momentum which had been gained was lost, first in England, then in other European countries and Japan. Although there had been considerable activity in Japan around 1960, the Japanese Government had not provided special research grants, and Japanese research was incomparably smaller than the US endeavours. As a consequence, even after the *ALPAC Report*, no drastic changes in Japanese efforts in machine translation took place. It was then well known that machine translation based solely on syntax was impossible and that the semantic aspects of language must be incorporated to some extent. Little had been accomplished in formal semantics, however, compared with the wealth of theoretical developments in syntax, and interest in machine translation research inevitably waned.

For these reasons, it is often stated that little research on machine translation was done, particularly in the West, between 1965 and the late 1970s. In truth, like the flow of underground streams, the intermittent efforts which were made during this period should not be overlooked. The Russian–English translation system which had been developed at Georgetown University (one of the leading centers for machine translation research in America) was moved in 1965 to the Rome Air Defense Center at Rome, New York. There the systems was improved and became a functional system. The translations which it provided required post-editing, but the system was capable of translating daily a considerable volume of Russian scientific literature. The fact that this system has been used routinely for more than two decades, with virtually no fundamental changes in its internal structure, is truly remarkable. It goes without saying that a wide range of knowledge, experience, and expertise on the practical problems involved in machine translation are contained in that system.

One of the early developers of the Georgetown University system was Peter Toma. Based upon his experiences at Georgetown, he was able to establish a private company which developed a new machine translation system, SYSTRAN. Completed in 1973, SYSTRAN was another Russian–English translating machine and was tested by a number of agencies and institutes in several small-scale projects. Thereafter, SYSTRAN was used for the development of English–French, French–English, English–Italian, and other systems. Moreover, beginning in 1976, the European Community started to use the English–French SYSTRAN system. Improvements were made over a period of several years, and it is still in use today. It is even reported that the Russian–English SYSTRAN system was used for the translation of technical documents relating to the Soyuz–Apollo docking experiment. In addition to SYSTRAN, several other machine translation systems were developed in the 1970s, one of which was the LOGOS system for use in translating from English to Vietnamese during the Vietnam war.

In Europe, relatively little work on machine translation was done during this time, but, although reduced in size, the Automatic Translation Research Institute associated with the French CNRS has been continuously engaged in such work up to the present. Particularly noteworthy have been the sustained efforts of the center's Chief, Bernard Vauquois, to continue with basic research on machine translation. While the number of personnel gradually dropped from several score during its period of peak activity to less than twenty today, work on Russian–French, French–Russian, and other translation systems has continued. It is worth noting that Vauquois greatly valued the research efforts of Japanese visitors and worked on several long-term projects in collaboration with Japanese teams. More recently, the center has been involved in the development of a machine translation system which translates into the Malayan language.

Construction of dictionaries of technical terms

The problems created by the difficulties of translation among various natural languages have been the source of bitter experience for many centuries in Europe. As a result, many European universities have established courses or departments of translation, encouraged research in this field, and trained a large number of people for employment as translators and interpreters. The greatest headache for the translator is the appearance of specialist technical terms in the text to be translated. For example, it would be difficult even for a graduate in electronics to translate the Japanese term:

環境　電磁　工学

KANKYO DENJI KOGAKU

environment-electromagnetism-engineering

without specific knowledge of the corresponding term in English. It is therefore necessary to have a small armada of up-to-date dictionaries or lists of terms on technical subjects, and to look up such terms individually. The looking-up procedure, however, is costly in terms of time, and the efficiency of the translator therefore suffers greatly. Because of this, computer systems have been designed which contain many such technical terms and which are capable of providing the appropriate translation in a variety of languages.

In at least four organizations heavily involved in machine translation (the European Community, the German Translation Bureau, the Translation Department of Siemens AG, and the Translation Bureau of the Canadian Government), specialists are employed to collect technical terms and decide upon appropriate translations. On a daily basis, they gather new terms, provide a translation, and update the computer dictionary's technical database. In these organizations, electronic dictionaries now exist which contain as many as several hundred thousand or a million technical terms. When the translator is given a text to translate, he scans the entire text and underlines those technical terms which are unknown to him. The underlined text is then passed on to the individuals whose job is to input the new terms into the computer. The computer then provides translations of the new terms, based upon its current dictionary entry, and the results are given to the professional translator, who checks the computer translation and accepts it or edits it. This type of computer assistance provided to human translators is called 'computer-aided translation' and has been used for more than a decade at many locations in Europe and Canada.

The TAUM METEO system

There is one further machine translation project which should not be overlooked: the so-called TAUM METEO system, which is used in Canada to translate the text of weather forecasts from English into French. Since Canada is a country where both English and French are official national languages, both must be used for all governmental documents. Current machine translation capabilities are, however, such that it is not possible to perform all translation automatically. Fortunately, the texts of weather forecasts supplied by weather stations throughout Canada consist nearly entirely of standard phrases in comparatively short sentences. Canadian machine translation researchers were perceptive enough to see that this was an ideal area for the application of available machine translation techniques.

The development of this English–French weather forecast translation system was done at Montreal University (thus the acronym, TAUM: Traduction Automatique de l'Université de Montreal). Several times a day, weather forecasts arrive in English from meteorological stations throughout Canada, in the form of telex messages. At Montreal, these messages are fed directly into the TAUM METEO computer, translated immediately into French, and sent to newspaper offices, broadcasting stations, and other news outlets. This system began operation in 1978 and today works 24 hours a day.

The capabilities of this system are based upon the fact that sentences used in weather forecasting are generally short and frequently employ a standard phraseology. For example,

> Tomorrow will be clear or partly cloudy with
> a north-easterly wind

Such sentences often have a somewhat unusual structure and thus require rules of grammar applicable to them. Generally speaking, the linguistic expressions used in a specific field such as meteorology are referred to as a 'sublanguage', and rules of grammar are devised to work within the confines of that sublanguage. For this purpose it is not necessary to consider all of the rules of grammar which are valid for the entire language. On the contrary, in order to improve the quality of the translations within the specified domain, it is absolutely essential that rules of grammar be formulated which apply only to the sublanguage. Although machine translation systems have, as a matter of necessity, worked with a subset of natural languages, the concept of a sublanguage was first implemented by the Montreal researchers in the TAUM METEO system and has contributed centrally to the success of not only their own system but others as well.

It must be said, however, that the TAUM METEO system has not achieved automatic translation into French of any and all English sentences concerned with meteorology. When the weather varies within a normal range, the machine translations are sufficient, but when the forecast draws attention to unusual weather conditions, such as a severe cold front, complex sentences which the translation system cannot handle emerge and the translations fail. The TAUM METEO system is in fact configured such that it knows when it has succeeded and when it has failed in translation. When successful and not in need of checking by a human translator, it sends out the translated material directly. It can therefore be described as a completely automatic translation system. Those sentences which the computer fails to translate are displayed on the computer terminal display unit and the attending translator corrects the translation prior to output (Fig. 1.3).

Since those sentences which the computer fails to translate are, from the human point of view, not in fact terribly complex, the translator is normally able to translate the English sentence into French immediately. As I mentioned above, the system works 24 hours a day, thus requiring a night shift of human translators in attendance. When the system was first implemented, its success

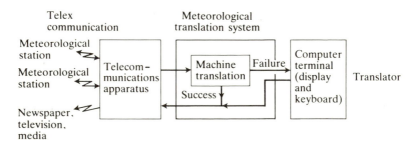

Fig. 1.3 Diagram of the TAUM METEO system, a machine translation system for translating weather reports from English into French.

rate was about 85 per cent, but as a result of recent improvements, successful automatic translations are now achieved with about 95 per cent of the input English sentences. This very high rate of success is due primarily to two factors: (i) The sentences to be translated all fall within an extremely narrow specialized field, and (ii) sentences which the computer system has failed to translate are handed over in their entirety to a human translator, thus ensuring that the system as a whole (including the human translator) will succeed. No machine translation system in the world is of greater practical value than the TAUM METEO system and, in this sense, the successes of TAUM METEO have encouraged and emboldened many researchers in the field of machine translation.

The successes of this translation system were closely observed by the Canadian Government, which then ordered the development of a second system, TAUM AVIATION, from the Montreal University Machine Translation Group. This system was originally intended for automatic translation of technical material relating to the maintenance of aeroplanes, but once it was recognized that too wide a range of topics was encompassed within that field, it was confined to the maintenance of 'pressurized oil systems'. Development proceeded for several years along two lines, grammar and dictionary data, but in the end the project ended in failure. It seems there simply were not two fish in that pond!

It is not entirely clear what the source of the failure was, but it is said that one cause of problems was the inability to select the appropriate terms in French from a number of possible translations of the English. Another source of difficulty was the fact that the costs of adding new terms into the system's dictionary became exceedingly high. It did not suffice simply to insert corresponding English and French terms: it was necessary to input a large amount of related information, including the associated articles, forms of usage, customary usage, and alternative meanings. Thus each expansion of the dictionary was expensive in both time and effort.

A return to fundamental research

In Japan, noteworthy research on machine translation itself was virtually non-existent in the years following 1968. Instead there was a return to issues of basic research. In my own laboratory, a software system was built which could automatically analyse sentences which had been rewritten using grammatical rules for sentence analysis that were designed expressly for ease of understanding by human readers. We also pursued research on the syntactic analysis of text, the abstraction of facts found in the text, and their transformation into a form which could be easily utilized by computer (that is, a kind of 'semantic network' system, which will be described in fuller detail below). The deduction of abbreviated terms, which are frequently used in Japanese text*, and methods for determining the referents of pronouns were also explored. In other words, we had returned to quite fundamental research questions, as indeed had most of the other Japanese researchers previously involved in machine translation research proper.

At the same time, there were dizzying developments in linguistic theory, starting with the publication in 1957 of Chomsky's *Syntactic Structures*,[5] and continuing until 1965 and the full development of his theory in *Aspects of the Theory of Syntax*.[6] During this period, many researchers within the Chomskian school published new theories, the significance of which should

---

*This is a problem peculiar to Japanese, in which lengthy proper nouns or technical terms are shortened by means of selective use of key ideograms, from within a string of 4–10 in the full term. Unlike an English abbreviation, for which the meaning of several letters will be completely ambiguous to someone unfamiliar with the abbreviation, the Japanese abbreviations retain a few of the essential semantic units and therefore convey some, if incomplete, meaning. Consequently, deduction of the full phrase is often possible. For example, the public telephone utility, Nippon Telegraph and Telephone Public Corporation, is written in full as:

日本 電信 電話 公社

NIPPON DENSHIN DENWA KOSHA

Japan Telegraph and Telephone Corporation

but is often abbreviated as:

電電公社

DENDEN KOSHA

which literally means 'electrical electrical corporation', but carries enough semantic information that deduction of the full name of the corporation is possible.

not be overlooked. Unfortunately, it was not an easy matter for those working on the engineering side of machine translation fully to understand, digest, and apply the multifarious developments in linguistic theory appearing at that time. These theories had not been distilled into a form usable in machine translation, and indeed, many of us did not immediately make use of them.

At about the same time, a new theory radically different from the ideas of the Chomsky school was also put forward: the so-called case grammar, developed by Charles Fillmore and systematized in 1968.[7] Case grammar is described more fully in Chapter 5, but briefly, it is a technique for describing the relationships of words in a sentence from a semantic perspective. Emphasis is placed on the verbs of the sentence, and questions are asked about what kinds of nouns can become the subject undertaking the action of that verb, what things (nouns) are the recipients (patients) of that action, and what are the relationships among the other nouns in the sentence (the relationship of the word to the main verb is called its 'case').

For example, the actor of a verb such as 'eat' is generally an animal, and the object or 'patient' of that action is 'something edible' or 'food'. These are relationships inherent in the meanings of these words. Using this type of description, let us examine a sentence such as

The meat was eaten by the cat.

Here, 'the meat' must be the object of the action, 'eating', despite the fact that 'meat' is grammatically the subject of the sentence.

Case grammar is particularly well suited for the analysis of the Japanese language. Since the sequence of phrases within a Japanese sentence can be freely changed, the order of such phrases reveal relatively little about the meaning of the sentence. Moreover, the suffixes following nouns in Japanese (WA, GA, WO, NI, MO, DE) are used in several meanings and cannot be relied upon to specify the grammatical case of nouns unambiguously.

When, for example, one wants to say 'I want to eat fish' in Japanese, the usage of WA, GA, or alternative suffixes is very complex.

私は魚が食べたい。
WATASHI WA* SAKANA GA TABETAI.

I fish want to eat

Using case-grammatical analysis, 'I' becomes the actor of the main verb 'eat' and, despite the fact that 'fish' is followed by the GA suffix denoting the grammatical subject, it is understood as the object of the verb 'to eat'. Since 1970, nearly all computer systems which analyse the Japanese language have used various implementations of case grammar.

In 1970, Terry Winograd of MIT constructed a computer system which allowed for free-flowing conversations between the computer and human beings. With that system he demonstrated that when handling words in a computer, knowledge concerning that part of the world with which the language is concerned is required, in addition to knowledge of syntax and semantics. This finding came as a shock to many linguists, for although it had long been known that such background knowledge is necessary, it had been thought that it would be impossible to build a coherent system if the realm of discourse was enlarged to that degree. As a consequence, linguists had purposely avoided the entire problem of the relationship between language and the world it described. Perhaps for the same reason and because there are ambiguous and strongly subjective factors involved in meaning, the problems of semantics within linguistics proper were largely avoided. In other words, linguistics up until this time was concerned solely with the most easily handled linguistic phenomena, such as the problems of vocabulary and the syntax of clearly defined sentences.

Winograd provided his computer with information concerning the placement of the building blocks which were the topic of his man–machine conversations. For example, he succeeded in having conversations of the following kind:

---

*One of the two principal uses of the particle WA is to indicate the thematic subject of a sentence; it is often translated 'as for' and is followed by a noun phrase. GA typically signifies the grammatical subject, although not in this example.

MAN: Find a block larger than the one which you now hold and put it in the box.
MACHINE: I understand.
MAN: What is in the box?
MACHINE: There are a blue, four-sided pyramid and a block.
MAN: Which object is closer?
MACHINE: The block.

This conversation may seem quite unremarkable, but it was an extremely complex computer program which was able to produce such factual statements. For example, the computer needed to know: (i) that 'the box', which was referred to twice, specified the same box in both instances; (ii) that there had previously been one object in the box; (iii) that the object which it put into the box was a block; and (iv) what the actual relationship of the objects was. Furthermore, when asked 'which' of the objects, the computer had to decide that the question referred to the immediately previous sentence in which the pyramid and the block were specified. Even if there were several similar objects, they were not what the 'which' of the human inquiry now referred to.

Winograd's contribution was to demonstrate the importance of using knowledge about the universe under discussion for the correct interpretation of the sentences in the conversation. This lesson also became the starting point for subsequent developments in the field of so-called conversational grammar, in which the relationships among sentences are dealt with.

Nevertheless, from the point of view of the development of functional machine translation systems, there were relatively few new developments, and the Dark Ages of machine translation continued. There were, however, many interesting developments with regard to the fundamental linguistics which must underlie any sophisticated machine translation system. Moreover, continued efforts were made to improve previously existing systems. Even in academic work and research there are fashions, and the fashions may change suddenly from lightness to darkness and back again. What is most important here is what efforts are sustained during a 'Dark Age' of academic unfashionability; an Age of Enlightenment cannot occur without a preceding Dark Age. The force which eventually can turn an era of darkness into one of light is the accumulation of many subtle effects. In fact, an era of Enlightenment may be regarded as one during which the fruits of slow and subtle development are harvested, but the harvest cannot be had without a long period of growth following the sowing of seeds.

2 Revival of machine translation

A new start

At the 18th International Meeting of the Information Processing Society of Japan, held in October 1977, I was invited to give a lecture on the 'Past, Present, and Future of Computational Linguistics', and there I took the opportunity to stress the fact that the time had come to undertake research on machine translation once again. Remarkable progress in computer hardware had been made since the 1960s and it was already possible to store and use electronic dictionaries containing many tens of thousands of words. Moreover, it was becoming possible for computer systems to handle Japanese ideograms, and the Japanese word processors which are today in such wide use were first produced in the autumn of 1978. Since the early 1960s there had been remarkable advances in linguistic theory, which had been formulated in such a way that they could be implemented directly in computer systems. In the late 1970s the TAUM METEO system had already been put into service, and the SYSTRAN machine translation system was being refined for testing within the European Community. Furthermore, world affairs were making translation ever more essential for maintenance of international relations. Developments in international travel and communication had further reduced the size of the world, and trade was becoming increasingly international. In the space of a few decades, the world had become a place where instantaneous transmission of all kinds of information was necessary.

There remained the need, however, to study in depth the linguistic problems inherent to machine translation. Despite the fact that a significant body of basic research on language processing by computers had been accumulated since the 1960s, the remaining linguistic problems were tremendous, and simple solutions could not be hoped for. Moreover, because of the *ALPAC Report* and its repercussions, researchers in machine translation had had the bitter experience of a complete halt of the funding needed to continue the research which had been begun in the admittedly somewhat over-zealous period of the 1950s and 1960s. There was consequently a clear realization that in the next era of machine translation research, working systems must be developed, and furthermore, that research results must be reliable and unambiguous.

A system for translating article titles

In this stringent atmosphere, the first project to be undertaken in Japan following the *ALPAC Report* was translation of the titles of scientific and technical papers from English into Japanese. In light of the successes of the TAUM METEO system, it was understood that the field within which translation was to be undertaken must be narrow. There was considerable interest at this point in time in the construction of scientific databases, due to the rapid developments in this field in America. This combination of interests led to the idea of a system for translating the titles of English-language scientific works and the automatic construction of a Japanese-language database containing that information. From a technical point of view, the most interesting development was the insight that the linguistic expressions in scientific titles are not of limitless variety, but rather include a relatively small number of fundamental structures, with emphasis on prepositional phrases.

Although there are several reasons for difficulties in the analysis of complete sentences, one major problem is that there is a virtually endless variety of sentence structures. Chomsky's idea of a phase structure grammar was one method for handling this tremendous variety, but it was apparent that the titles of technical papers could be dealt with using an even simpler set of rules. At this juncture, some 10 000 article titles were collected and analysed, and it was found that, excluding structures which included the repetition of nouns (for example, 'The current state of machine translation in Japan, America, France, . . .'), there were approximately 1000 kinds of linguistic structure used. By means of further simplifications and generalizations, made possible by automata theory, it was discovered that, surprisingly, there are only 18 fundamentally different structures used in such titles.

It was then understood that the structure of article titles could easily be analysed, but if the meaning of the words in such titles is not understood, the relationships of phrases within titles could not be accurately translated. For example, in a title such as

Understanding line drawings of scenes with shadows

it can be seen that, in terms of sentence structure, 'with shadows' could refer to 'understanding' or to 'scenes'. In the former case, the title means:

Understanding by means of the shadows in line drawings
of scenes

whereas, in the latter case, the title means:

Understanding line drawings of scenes in which shadows
are present

Because we know that there are, semantically, many relations betweens 'scenes' and 'shadows', the latter interpretation is more likely to be the correct one. Clearly, in order to be able to choose the more realistic translation in cases such as this, a minimum of semantic information must be introduced. Since it is well known that a quagmire of difficulties is encountered when any attempt is made to undertake complete treatment of semantic problems, it was decided to limit semantic analysis to the minimum required for successful translation.

The system which was developed along these lines between 1978 and 1979 was quite successful, given its relative simplicity. Even in the case of rather complex English titles, the system was capable of translating into Japanese, such that at least a rough understanding of the title was possible. Several examples of the output are shown in Fig. 2.1. Fortunately, the system was introduced at the Tsukuba Computer Center of the Industrial Technology Institute, where it was used extensively and so became widely known.

It must be said, however, that this system was a simple one designed for use in a limited field, and there were naturally a considerable number of deficiencies in it. For example, an (extremely unlikely) title such as 'He is a boy' would be translated into Japanese as 'Helium is a boy', for the reason that the system had been designed for technical translations and 'He' was

Superquadrics and angle-preserving transformations

超二次関数と角度保存変換

CHONIJIKANSU TO KAKUDO HOZON
HENKAN

Computer graphics display hardware

コンピュータグラフィクス表示ハードウェア

Computer graphics HYOJI hardware

A comparison of antialiasing techniques

Antialiasing 技術の比較

Antialiasing GIJUTSU NO HIKAKU

Homogenous coordinates and projective planes in computer
graphics

コンピュータグラフィクスでの均一座標と射影平面

Computer graphics DE NO KIN'ITSU ZAHYO TO
SHA'EI HEIMEN

Computer graphics and the business executive – the new
management team

コンピュータグラフィクスと企業経営者 ── 新しい管理チーム

Computer graphics TO KIGYOKEIEISHA –
ATARASHII KANRI team

A guide to sources of information about computer graphics

コンピュータグラフィクスについて情報源へのガイド

Computer graphics NI TSUITE JOHOGEN E NO guide

Fig. 2.1 Results of English-to-Japanese translation of the titles of
journal articles.

recorded in the dictionary only as the element helium and not as the personal pronoun. With only a small number of exceptions, the terms which had been included in the dictionary all had unique translations — that is, one English word for each Japanese word — and there were no functions for checking the overall semantic cohesion of the translated sentence.

Well aware of the strengths and weaknesses of this system, we developed a second system for translating the titles of Japanese technical papers into English, and a similar system for translation into French. One interesting insight obtained during these projects was that it is essential to have involved in the developmental work researchers who are native speakers of the target language. Specifically, a comparison of the Japanese–English and Japanese–French translations showed that the quality of the translations was better in the French system. The systems had identical portions involved in the analysis of the Japanese, but the part of the system which generated English-language phrases had been developed by Japanese researchers, whereas that for the French generation portion was done with the help of J. Hubert, a champion 'Go' player in France, a grade four player in Japan, and a visiting researcher in our department. Examples of the French system are shown in Fig. 2.2.

The current situation in Japan

Subsequently, the Japanese Government recognized the importance of machine translation and decided to support such research, to overcome language difficulties encountered in the rapid transmission of technical information at scientific conferences and so on. A four-year plan for research was thus recommended in 1982. At the same time, private businesses gradually began to show interest in machine translation, due to the fact that companies involved in the export of electronic and computer equipment spend large amounts on the translation of user manuals and instruction booklets. For example, for the export of a single mainframe computer, several hundred thousand pages must be translated into English — a pile of technical material extending from the floor to the ceiling! Translation costs alone can therefore run to several hundred thousand dollars.

According to a survey undertaken in 1980 by the Machine

官庁統計における電子計算機の問題
KANCHO TOKEI NI OKERU DENSHI
KEISANKI NO MONDAI

official statistics concerning calculator of problem

Probleme d'ordinateur dans la statistique de bureau de
l'administration

初等的な論理学の定理の証明のための
非自己発明的なプログラム

SHOTOTEKI NA RONRIGAKU NO TEIRI NO
SHOMEI NO TAME NO HIJIKI
HATSUMEITEKI NA program

elementary of logic of theorem of proof for
non-heuristic of program

Programme non-heuristique pour la demonstration de

theoreme de logique elementaire

格子模様の重ね合せで写真フィルムに
デイジタル・データを記録するシステム

KOSHI MOYO NO KASANEAWASE DE SHASHIN
film NI digital/data WO KIROKU SURU system

lattice pattern of super-position photographic film to
digital/data record system

Systeme enregistrant les donnees numeriques pour le film
photographique par superposition de motif en treillis

Fig. 2.2 Examples of the output of the Japanese-to-French title-translating system (developed in collaboration with J. Hubert).

Translation Research Committee of the Japanese Electronics Society, at least ten million pages of technical documents are translated yearly in Japan. If each page costs 2000 yen (10 dollars), then some 20 billion yen (100 million dollars) is spent on translation in Japan alone. The need to reduce the costs of technical translation is thus apparent.

In these circumstances, it was inevitable that research on machine translation would be pursued at the research institutes of various Japanese computer makers, and since 1984 various commercial systems have been introduced. Today, more than ten corporations have machine translation systems on the market. This commercial activity can be seen not merely as an enthusiasm for producing commercial products, but also as a manifestation of the development of techniques for the handling of words by computers, in a world which is becoming increasingly concerned with information.

The commercial machine translation systems vary widely, differing in their quality and translation features. With the large-scale mainframe translation systems, which contain large dictionaries and rapid translation capabilities, it is entirely possible to develop translation services in which individual users communicate with the mainframe by normal telephone lines and receive translated text in return. Although less satisfactory results can be expected from translation software used with personal computers, simplified systems in which the user can rewrite the original document in a less complex style, or in which a user with a good knowledge of the target language can undertake the necessary post-editing, are already feasible. In the case of either small or large machine translation systems, automatic translation is far from perfect, but rapid developments can be expected under the pressures of user demands and competition among businesses. Undoubtedly, users will learn the idiosyncracies and limitations of such systems and be able to work around them. The introduction of such systems into the market-place is thus extremely important and will encourage developments in new applications.

Developments in other countries

The problems of translation are most acutely felt in the European Community, where, with the addition of Spain and Portugal in 1986, there are twelve nations and nine recognized languages. It therefore follows that translations must be done among all of these languages — French into English, English into French, German into English, and so on — giving a total of 72 (8×9) directions of translation! In 1985, some 2000 translators were employed to make the translations among the seven languages which were recognized at that time (42 directions of translation), and it is projected that more than 3000 translators will be needed for the translations among the present nine languages. Needless to say, the salaries of 2000 specialist translators already amounted to a considerable sum, but those for 3000 translators are seen as a serious problem.

Another problem arises with regard to the time required for translation. For example, a European Community meeting cannot be officially opened until the Proceedings have been translated into all the officially recognized languages. For this reason, the entire conference is necessarily delayed by the translation process, and this temporal lag is a problem which plagues the activity of the European Community.

In the light of these issues, the Community decided to employ the SYSTRAN English–French machine translation system, starting in 1976, and efforts were made to improve the system to be of practical use. The system was put into use in 1981 for the translation of certain documents known to use relatively simple linguistic style. Several other systems have been developed within SYSTRAN, such as German–English, English–German, and English–Italian systems. It was, however, soon found that a single computer cannot easily handle a variety of SYSTRAN systems, a fact which then led to interest in the possibility of constructing multilingual translation systems. In other words, a system was imagined in which, once the analysis of a sentence in one language had been accomplished, it would then be translated into any one of several languages. This is the situation illustrated in Fig. 2.3.

At the request of the European Community, the possibility of a multilingual translation system was studied by researchers in various European universities for several years, starting in 1978. It was decided that the first period of active research and development should last for five-and-a-half years, beginning in 1983. That research project has been named the EUROTRA plan. A second five-year project is envisaged to follow the first research period, based on the results obtained; during this the development of practicable systems are anticipated. Because it was thought that similar goals could be accomplished at a single geographical location only with considerable difficulties, the pro-

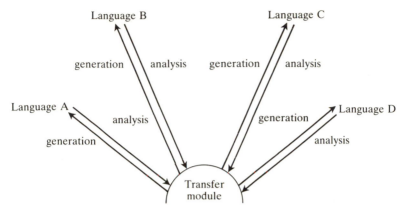

Fig. 2.3 Schematic diagram of a multilingual translation system. This is the basic idea behind the EUROTRA plan, in which the central module for transfer operations is kept to a minimum size, and combined with modules for the analysis and generation of sentences in the individual languages. In a system requiring translation between English and Japanese, the module for transfer operations must be greatly increased in size.

ject is to be undertaken in each of the nations of the European Community on a collaborative basis. It will most probably make use of a computer network system extending throughout Europe, but details of the communications system have not been finalized. This is a truly ambitious project and is of extreme interest, not only regarding the outcome of the research *per se* but also in terms of the success of the communications network.

In America the influence of the *ALPAC Report* remains considerable even today and, 20 years on, little direct research on machine translation is being done. It is also true, of course, that English has become the principal international language, thus draining some of the motivation for machine translation research in the USA. Noteworthy, however, is the fact that a wide range of technical by-products has been realized as a direct result of previous US research on machine translation. These include various interactive systems which allow for human–computer interaction, and techniques for natural language processing. Thus, while giving due credit to the considerable efforts in machine translation which were made in earlier years, it must be said that the machine translation techniques currently in use in the USA are essentially those of the mid-1960s, or slight improvements on them undertaken by commercial organizations. Compared with developments in Japan and Europe, those in the USA have been slow.

The development of a database system containing a thorough dictionary as an aid for human translators was first begun in the early 1960s. Thereafter, due to the independent efforts of the West German Translation Bureau, the Translation Department of the Siemens Corporation, the European Community, and the Canadian Government, the development of such dictionaries has progressed a long way, and they have been of considerable use in actual translation work. In recent years, however, there have been further developments, principally in two directions. The first has been an effort to include the linguistic idiosyncracies of as many different languages as possible in the dictionary databases. The second has been the attempt to fit as many languages as possible into a single database.

Finally, it should be noted that, stimulated in part by the research on machine translation in Japan and Europe, there has recently been a re-evaluation of machine translation in America. The resurgence of research activity can therefore be said to be world-wide, and not confined solely to Japan. It is, however, unwise to harbour expectations which are too great, and researchers have the responsibility of building reliable machine translation systems which will be of genuine practical value and which will meet at least some of the demands and expectations of society in general. In other words, care must be taken not to allow the occurrence of another *ALPAC Report*. In this light, it is essential that researchers have in mind modest and realistic targets which can be attained with certainty.

3 A perspective on machine translation

What is translation?

Perhaps the first researcher to address this question seriously and to produce some practical results was E. Nida. While engaged in translating the Bible, he ruminated on the problems inherent in producing translations. His first insight was simply that there are several levels at which translations can be made. The first is based upon traditional linguistics and entails the 'mapping' of the words and grammatical structures of one language (the 'source' language) on to those of a second language (the 'target' language). The second level is based upon communication theory, and involves the construction of sentences in the target language which have the same meaning as expressed in the source language, regardless of the degree to which the translation retains the grammatical structures of the original text. The third level is based upon sociolinguistics. That is, a sentence which is expressed within a given culture will be understood and lead to specific behaviour within that culture; the proper translation of a sentence at the third level should therefore lead to the identical behaviour within a different culture. This view of translation is an extremely sophisticated one, in which consideration must be given to the results produced in light of the relationship between the sentence and the cultural setting.

Stated more simply, these three views can be summarized as translation based upon syntax, translation based upon semantics, and translation based upon pragmatics (social results). Needless to say, given the nature of Nida's practical work in translating the Christian Bible, he was most concerned with the third level of translation. He believed that when dealing with works such as the Bible, translation must be done quite irrespective of the structure and syntax of the original work, so that the semantic content and affective weight of the original text is maintained in the translation. In practice, many difficulties arise in this kind of translation.

One such problem is the existence of words for which there are concepts in one society, but neither words nor concepts in another. When such a concept is used in the speech of the first society, translation into the speech of the second society is quite impossible, no matter how 'semantic' a translation is attempted. When such difficulties are encountered, it is difficult to know if any solution will allow for a sufficiently close concept to be brought to mind by the reader or listener in the target language. One mundane example can be seen in the Japanese phrase:

腰が痛い。

KOSHI GA ITAI.

hip hurts

[My] hip hurts.

which is surprisingly difficult to translate into English. Neither 'waist' nor 'hip' is what is meant by the Japanese KOSHI, and as a result KOSHI GA ITAI is often mistranslated as 'I have a pain in the back'.

According to the linguist Tetsuya Kunihiro, Japanese words used in cooking, such as 'yaku' and 'niru', correspond imprecisely to their English translations (Fig. 3.1).[9] In other words, there is a lack of isomorphism between the normally used translations of the Japanese and English words in cooking. Although cooking practices in English-speaking and Japanese-speaking societies are close enough that the concepts are familiar to both, the distinctions drawn between words are slightly different.

Nida's ideas about the most sophisticated level of translation might be rephrased as follows: Translation is the process of transmitting the meaning, feeling, and artistic value of speech completely, so that it is fully understandable by the society of the target language. In order to accomplish this, the sentences in the source language must first be fully understood, then transformed to fit the world of understanding of the society of the target language, and finally reconstructed as sentences in the world of understanding of the target language. Again, it is the transformation and transmission of the ways of thinking of one culture into

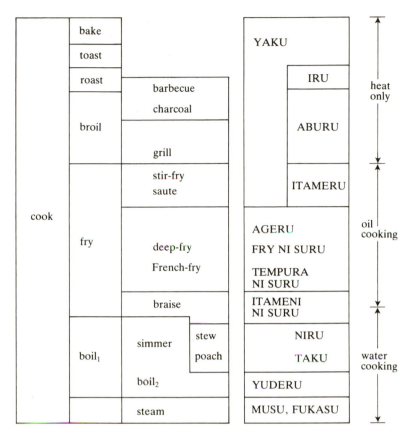

Fig. 3.1 A comparison of the cooking verbs used in Japanese and English (from Kunihiro, 1981).

those of another. Certainly, the best translations of literary works achieve this high level of intercultural translation, but in order to approach such an ideal, understanding of the language (as distinct from its literal decoding) is needed. Currently, computers come nowhere near such a level of translation. At best, current efforts in machine translation are aimed at Nida's first two levels of translation. Although research efforts have also been made on the understanding of sentences, as in Nida's third level, it is simply not known what is indicated when we say that 'a human being has attained an understanding' of some segment of language.

Varieties of translation texts

Translation can be considered on the basis of a classification of the different types of text that are to undergo transformation into a foreign language. The following four categories can be distinguished:

(i) Poetry and literary works, whose value lies primarily in the emotion conveyed to the reader.

(ii) Legal documents and contracts, in which the interpretation of the sentences lies primarily in its unambiguous, logical effects.

(iii) Technical and scientific material, in which correct understanding by specialists within a limited field is sufficient.

(iv) Article titles and sentences for use in information retrieval systems, in which it is sufficient to communicate to the reader whether or not information of interest can be found in the relevant article.

An alternative categorization can also be imagined:

(i) Translation into a transitory stage or into a final stage.

(ii) Translation for internal use or for a wide audience.

(iii) Translation in which detailed accuracy is required or in which a general outline suffices.

(iv) Translation for subsequent publication or not.

It is essential to consider where machine translation stands in such classifications and perspectives. To obtain a detached, objective stance on the current capabilities and future possibilities of machine translation, excessively high expectations and pessimistic underevaluation must both be avoided.

Current machine translation is not accomplished by means of translation based upon *understanding* of the written text. Similarly, it is not yet possible to obtain an interpretation of a sentence in relation to world knowledge or the conditions under which the sentences were uttered — facts which a human translator would normally use for the selection of the most suitable expression in the target language. Since computers do not achieve 'understanding' and do not translate in the light of 'world knowledge', the suspicion arises that machine translation may be fundamentally insufficient, and that in comparison with human translations, machine translation can hardly be considered 'translation'. However, even with machine translation there are cases of translation which are successful at the highest of Nida's levels. Such cases can be found when the sentences to be translated are confined to technical and scientific matter and deal with topics from very narrow fields.

For example, consider a manual concerned with the structure of an electric refrigerator, in which there is the following description requiring translation:

> This electric refrigerator consists of three compartments. On the inner surface of the door is a place for storing bottles....

Since this is the description of a specific type of equipment, there is virtually no need for a 'semantic' translation in which it would be necessary to change the structure of the sentences into a completely different form due to social or cultural differences. In other words, it is likely that the results will be the same, whether or not the sociolinguistic perspective is considered. Furthermore, if the sentences to be translated are written as accurately and precisely as possible, there may well be no semantic ambiguities, given the limited field of the work. There should therefore be no difficulty in understanding the meaning of the sentences which have been translated at the level of Nida's first category of translation (by means of a procedure for transferring the terms and structures of one language into another). The full purpose of the translation will therefore be achieved.

Thus, machine translations should work with sentences of this particular kind, whereas sentences which include more complex semantic factors are not suitable for current automatic translation systems. Stated more positively, sentences which contain complex semantic elements should be considered topics for future research.

It is essential that those who consider machine translation to be fundamentally incapable of high-quality work and those who are hostile to the research effort in machine translation should understand this cautious and realistic perspective, which is shared by most people working in the field. At the same time, those who are engaged in the research behind machine translation must fully appreciate the semantic limitations of current systems, and explain current capabilities and future prospects to the public at large in the light of those strengths and weaknesses.

Facts — speaker — environment

We have seen what kinds of sentences are suitable for machine translation. Let us now approach this topic from a slightly different perspective. Any sentence can be seen to contain the three elements illustrated in Fig. 3.2. The first element is the content or factual reality which the sentence describes. The second is the attitude or intentions of the speaker or writer; and the third is the situation in which the sentence was expressed, or its social context.

A sentence such as 'He is a young man' conveys the fact that there is a person who is male and young. On the other hand, the sentence 'He is likely to be a young man' conveys the same facts, but is expressed as a supposition. In other words, the second sentence communicates information concerning two different dimensions of reality: both a factual reality and the speaker's attitude or uncertainty with regard to that reality. As an example of a sentence containing the third level, consider simply 'I am a boy' or 'I am a girl'. Depending on the social factors and the emphasis placed on these words, various translations are possible — emphasizing the pronoun, the verb, or the noun. This would be rendered in Japanese as:

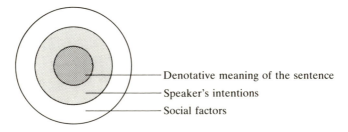

Fig. 3.2   Varieties of information contained in a sentence.

僕は少年だ。

BOKU WA SHONEN *DA*.

I young man *am*.

I am a boy.

俺は子供だ。

ORE WA *KODOMO* DA.

I child am.

I am a *child*.

私は少女です。

WATASHI WA SHOJO *DESU*.

I girl am.

I am a girl.

Each has a slightly different meaning which depends upon the social context of the sentence (and therefore the selection of pronouns, the politeness of the verb form, and the stress within the sentence itself).

Needless to say, machine translation can most easily handle the first of these three levels of translation. The second case, in which the speaker's attitude must also be considered for a proper translation, is gradually falling within the capacities of machine translation systems, but difficulties remain for complex expressions reflecting the speaker's psychological state. The third level remains a topic for future research.

In this sense, it is possible to consider technical and scientific topics, which deal with factual relationships and which, as a rule, have no social or cultural implications, as appropriate material for translation by current automatic systems. Even if the scope of the language is limited to a small subset of technical material, however, present day computers still cannot handle an extremely large number of linguistic phenomena. And even when the sentences are of a technical-scientific nature, there are of course cases when the translation is affected by social and cultural factors. Again, a mundane example might illustrate the point:

Change the Grade XX axle grease at monthly intervals.

Now it may be that Grade XX axle grease is unobtainable in the society into whose language the translation is to be done, so that Grade XX axle grease must be replaced with a corresponding axle grease, Grade XXX, in the translated sentence. Moreover, if Grade XXX axle grease needs replacement at half-monthly intervals, then the translation should be altered accordingly.

It is true that this is not strictly a problem in the actual technique of translation, but rather due to other factors. Nevertheless, in reality and in a commercial translation service, attention must be paid to such details. Since it is entirely conceivable that an accident could arise due to a literally accurate but nonetheless imperfect translation, it is extremely important to provide the appropriate information to the reader of the translated material. It simply cannot be glibly concluded that since translation work is done entirely on paper, no dangers can arise from it. There have in fact been cases of mistranslation which have caused sizeable accidents, and it is impossible to know in what way mistranslations could have social consequences.

Automatic translation of sentences

It is well-known that human beings commit mistranslations, and so do translating machines. The kinds of errors made by man and machine, however, differ in a variety of ways. In the case of human translators, translation is simply impossible if an understanding of the original work has not been achieved. For example, when simultaneous interpreters fail to grasp the meaning of spoken sentences, no translation is attempted and repetition of the speech is required. Even when a partial understanding of sentence fragments is achieved, if the overall meaning is not grasped, interpreters will be reluctant to offer any translation. This is true not only in simultaneous interpreting, but also in the translation of written sentences.

When a translator encounters a sentence which he understands partially but not completely, a strategy which is commonly employed is — like a machine translation system — to rely solely on the sentence structure and to produce a mechanical translation of the words in the appropriate order. An ordinary reader of such a translation would naturally fail to understand the meaning of the sentence, but a specialist with knowledge concerning the topics in the text (and particularly one with some understanding of the idiosyncracies of machine translation) would be able to obtain some degree of understanding of the content of the sentence and make informed guesses about the original meaning. It is therefore unwise to assert categorically that translation is impossible if the translator fails to understand the original work fully. In so far as the understanding of sentences depends upon the knowledge and wisdom of the final reader, a certain degree of mistranslation can be tolerated both in human and machine translations. Phrased differently, words are, in the final analysis, vehicles for transmitting information; if the information is conveyed, the translation is successful.

Translations of varying quality appear in the output from machine translation systems. These include:

(i) properly formed sentences,

(ii) properly formed sentences which include grammatical errors,

(iii) sentences which are difficult to read unless certain portions are rearranged,

(iv) sentences which are partly correct, but partly incorrect,

(v) sentences in which the word order is incorrect and a properly formed sentence is not produced,

(vi) only fragmented translation with omissions, and

(vii) no translated output.

Because of the variety of such translations, evaluation of the degree to which translation has been accomplished is extremely difficult. 'How well does that machine translate?' is consequently quite a profound query! And there are many who, having heard of the completion of a machine translation system, misunderstand this to mean that, regardless of the input, 100 per cent accurate translations are possible. But perhaps worse than such an expectation, some people appear to be satisfied with a numerical answer such as 70 per cent in answer to a question about how well an automatic system can translate! As discussed above, there is a variety of factors which contribute to the achievements and deficiencies of machine translation, and any evaluation in terms of percentages is unlikely to convey that complexity.

One more topic of importance in evaluating machine translation systems concerns the complexity of the sentences which the computer is given to translate. If it is presented with 'I am a boy',

'This is a desk', and other simple declarative sentences, successful translation will be achieved in well over 90 per cent of cases, but what would happen when it is given a typical article from *The New York Times* or a book on philosophy? In the cases of translation of Japanese texts, where subjects and objects are often omitted, and of extremely long and complex sentences, it is inevitable that the translated output will suffer considerably. It must therefore be concluded that the evaluation of machine translation systems is itself an extremely difficult matter when full consideration is given to all such factors.

As far back as the *ALPAC Report*, there have been suggestions for ways in which to evaluate the quality of translated works. The ALPAC method was devised in the early 1960s, but it remains of value today. Here, I will describe the method we use. It is relatively simple, and moreover, can easily be put into practice.

The first stage is an evaluation of the 'ease of understanding' of the translation and is simply the degree to which the text can be comprehended in a normal reading. In the case of translation from Japanese to English, the quality of the translation can be tested by investigating the degree to which the content is understood by a typical native speaker of English reading the translated text without seeing the Japanese original.

The second criterion tests 'fidelity': the degree to which the translated text faithfully reproduces the information contained in the original. This must be carried out by a professional translator who understands both languages, to determine the degree to which there are differences between the original and the translated text.

Since such tests of the quality of translation have often been based upon extremely subjective factors, it is essential to decide upon several detailed criteria for the evaluation and then for the evaluation to be performed strictly on the basis of those criteria. An example of a set of criteria is presented in Table 3.1.

Needless to say, a variety of other ways to evaluate the quality of translations comes to mind. For example, one well-used method is to evaluate the translation on the basis of the number of corrections which the translated text ultimately requires. Since, however, shorter sentences are normally more easily translated than longer sentences, such evaluation cannot be applied to

Table 3.1 Examples of criteria for evaluating translations.

Ease of understanding

---

1. The sentence has a clear meaning, with no possibility of misunderstanding. Appropriate grammar, terminology, and sentence structure are employed, and post-editing is not necessary.
2. The meaning of the sentence is clear, but there are slight problems in the sentence structure, grammar, or terminology. Post-editing would easily be accomplished by human hand.
3. The overall meaning can be grasped, but there are small ambiguities due to grammar or terminology. Post-editing would best be done with reference to the original work.
4. The quality of the translation is poor, and there are many problems of grammar and terminology. Some understanding of the sentence can be inferred after lengthy study of the translation, but it would be better to undertake the translation anew by human translator than to post-edit the machine translation.
5. No understanding of the contents of the translated sentence is possible. Human translation is required.

Fidelity

---

1. The entire contents of the original text have been translated into an easily readable form.
2. The entire contents of the original text have been faithfully reproduced and is easily understood in the translation, but minor post-editing is required.
3. The translation is generally faithful to the original, but some reorganization of the word-order is required.
4. The translation is generally faithful to the original text, but there are errors in the relationships between phrases, tenses, singular/plural, position of adverbs, etc. Post-editing requires structural transformations.
5. The structure and contents of the original work are not well preserved in translation, with portions of the original text incorrectly translated or the referents of words and phrases incorrect.
6. The structure and contents of the original are poorly translated. Phrases and clauses are missing, but valid sentence structure is achieved.
7. The translation reflects neither the structure nor the contents of the original text. Valid sentence structure is not achieved because of missing subjects, predicates, etc.

---

each sentence alone. Instead, evaluations must be made on the basis of the number of corrections divided by the number of words in the sentence. This technique gives an evaluation of the quality of translation per word and avoids a bias against longer sentences. Another criterion might be the speed with which the translated text can be read with understanding. Various other means for measuring the cohesion and user-acceptability of translations might also be devised.

The conventional technique used for commercial translation involves two steps: a novice or non-specialist translator produces a first draft, and then an experienced translator makes corrections to the translation. Within the European Community, translations are produced by teams of translators in which there are generally about three less experienced translators for every expert, who is responsible for making corrections and deciding upon the final version. Since this team-work approach is required even in human translations, it is overoptimistic to believe that the translated output from a computer can be given, as is, to the end-user. On the contrary, post-editing by a human translator is the more likely situation for most machine translation systems for years to come. Thus it is important that work on machine translation should today be aimed at increasing the number of translated sentences which are of such a quality that they can be easily post-edited.

Having established criteria for evaluating translated texts, the final evaluation will nevertheless differ considerably depending upon the intended use and the quality of translation which can be accepted. If one considers the case of the external third party reading the translated text, even if the translation is as good as category 1 in Table 3.1 (where essentially no problems are found), it is still necessary to have someone read the translation and confirm that errors are not present. If, on the other hand, one is interested in knowing the general contents of technical-scientific documents, translations which are in categories 1, 2, or 3 — that is, translations which are fairly easy to understand and generally accurate — can be used without the need for post-editing. For translations which are in any case going to receive post-editing by a professional translator, translations which are at least in category 3 in terms of ease of understanding and at least in category 4 in terms of accuracy should suffice. If, however, the

translation is of a still poorer quality, the entire translation process will undoubtedly need to be redone without the aid of the
machine translation system. In summary, the time and effort
which must be spent on the post-editing of the machine translation output has a large influence on the cost–performance of the
entire system.

It is essential to apply similar criteria to the original work to
determine how understandable and clearly written the source text
is prior to translation. Finally, a comparison of the quality of the
translation made by a professional human translator and that
made by a machine translation system must also be carried out.
In difficult specialized fields especially, human translators often
fail to achieve a full understanding of individual sentences and
end up performing mechanical translations based solely upon the
syntactic relationships within the sentence structure. In such
cases, human translators make many of the same mistakes that
machines do.

4 The process of machine translation

The principle of compositionality

In this chapter, the basic mechanisms by which machine translation takes place will be explained, starting at a simple level and gradually working to more complex levels.

The Japanese translation for 'good morning', OHAYO, is equivalent to the English because it is conceptually and functionally the same as the English term. Little explanation is necessary and, at least for the purposes of machine translation, no further rationale is required for such a translation beyond the fact that the English and Japanese terms are conceptually identical. Similar comments could be made with regard to the translation of most other isolated words: 'boy' translates as SHONEN, 'girl' translates as SHOJO, etc. The translation procedure in these cases requires no more than the use of a dictionary of terms, but there are cases in which a term in one language might have two or more possible translations: KYOTO DAIGAKU translates as 'Kyoto University', but TOKYO DAIGAKU translates as the 'University of Tokyo'. When referring to the US Government, KAIN translates as 'The House of Representatives', whereas in Britain it translates as 'The House of Commons'.

It is thus evident that computer systems must be made to store compound terms with their compound translations. As sentences become more complex, however, it becomes impossible to have the system remember the concepts of every possible phrase for one-to-one, unique translation. There is already a limitless number of combinations of adjectives and nouns (beautiful flower, pretty flower, beautiful girl, . . .), and the capacities of computers to store all such combinations as separate entities is soon exceeded. For this reason, compound terms are broken down into their more fundamental word units and a one-to-one translation is made at that level. This is the essence of the principle of compositionality. The phrase is then reconstructed in the target language (UTSUKUSHII HANA, KAWAII HANA, UTSU-KUSHII SHOJO, . . .) from a combination of terms.

When constructing a phrase consisting of several words, the word-order will not always be the same in the different languages. For this reason, it is necessary to consider the attributional or referential relationships among the words. A phrase such as 'very interesting story' consists of an adverb plus an adjective plus a noun. We are immediately cognizant that the adverb has the adjective as its referent and the adjective has the noun as its referent. If this phrase were to be translated into Japanese, first the individual words would be replaced by their Japanese counterparts and then the inter-relations of these parts of speech would be converted into the world of the Japanese language. In this example, since the parts of speech happen to be the same in Japanese and English, the word order is also the same. The translation is therefore:

非常に面白い話

HIJO NI OMOSHIROI HANASHI

very interesting story

If we were to translate this into French, however, the parts of speech differ from those in English and Japanese, and there are many cases in which an adjective–noun combination in English will be transformed into a noun–adjective combination in French. Since the adverb–adjective relation remains the same, the translation in French becomes 'histoire tres interessante'.

When reconstructing a single phrase, it is necessary to go through a procedure by which the original phrase is broken into smaller units, and the word-order of the entire translated phrase considered. When there are yet smaller units, this procedure must be repeated until the level of the smallest units, that is, words, is reached. Consequently, to translate such phrases, not only is a dictionary with corresponding words in the two languages required, but grammatical rules for the word-order transformations between the languages must also be provided. For example, a machine translation system would require rules such as:

> (Japanese) Adjective + noun = adjective + noun
> (English, some French, etc.)
> (Japanese) Adjective + noun = noun + adjective
> (some French, etc.)
> (Japanese) Adverb + adjective = adverb + adjective
> (English, French, etc.)

By means of the combined use of a word dictionary and grammatical rules of this kind, we can avoid the problem of having to enter into the computer all of the virtually infinite possibilities for compound words and their corresponding translations.

Current machine translation systems are built upon this principle of compositionality, thus greatly reducing the storage requirements of the systems. The translated sentence is therefore constructed from phrases in which the words correspond to the words in the original work. Those phrases which cannot be dealt with simply in terms of the principle of compositionality are treated as idioms. For example, the Japanese translation for 'good morning' is not rendered as YOI ASA (literally, 'good' + 'morning'), but as OHAYO (literally, 'earliness').

It is of interest, however, that high-quality human translations are *not* produced solely on the basis of the principle of compositionality. On the contrary, translations are done in consideration of the need to produce the same effects in the reader of one language as in the other. An example of English and Japanese with the same effects is as follows:

Here is a cup of tea for you.

which is properly translated as:

お茶をどうぞ
OCHA WO DOZO

tea please

A literal translation into the Japanese:

貴方のためにここにお茶が一杯あります。
ANATA NO TAME NI KOKO NI OCHA GA
IPPAI ARIMASU.

You for here tea   one cup is

is grammatically correct, but sounds distinctly strange to the Japanese ear.

There have been suggestions that, in view of the mammoth storage capacities of modern computers, all idiomatic expressions might simply be stored in the computer memory and used in translation. Unfortunately, there is a limitless number of such expressions, and no matter how large the storage capacities of computers, the actual handling, searching for, and comparing of such idioms would be unwieldy. Moreover, the usage of such well-translated idioms can change, depending upon the circumstances. Although OHAYO will be most appropriate in many cases, in others the more polite expression OHAYO GOZAIMASU might be the appropriate translation for 'good morning'.

Current machine translation systems are virtually powerless with regard to such problems, and it has proven difficult to extract workable principles of compositionality which are generally applicable. Therefore, to obtain the best translation possible, efforts have been devoted to the simultaneous treatment of the maximum number of adjacent words. As I discussed briefly at the start of Chapter 3, the reason why technical-scientific papers and other texts which deal with factual, objective relations are the principal target for machine translation efforts is that the establishment of principles of compositionality for such texts has been relatively easy.

Sentence analysis and translation

The process of translation is more complex for sentences than when dealing with phrases. For example, the sentence:

I would like an apple

can be interpreted in terms of grammatical rules, as follows:

Sentence = pronoun + verb + noun phrase
Verb phrase = auxiliary verb + verb
Noun phrase = article + noun

For translation into Japanese, the following interlingual transformation rules must be used:

(English) Pronoun + verb phrase + noun phrase $\Rightarrow$
(Japanese) Pronoun + noun phrase + verb phrase

which gives:

WATASHI WA HITOTSU RINGO WO
HOSHII DESU.
私は一つりんごをほしいです。

It is normally possible to omit 'one', so that the final translation would be:

私はりんごがほしいです。
WATASHI WA RINGO GA HOSHII DESU

or possibly:

私はりんごがほしい。
WATASHI WA RINGO GA HOSHII.

This translation process can be illustrated pictorially as in Fig. 4. 1. In the figure, the boxes contain strings of different parts of speech. When the interlingual transformation occurs, the order of these parts of speech can change and elements in the boxes may be removed or added.

It is well known that there are five fundamental sentence structures in English. If the subject of a sentence is designated S, the verb V, complements C, and direct and indirect objects O_2 and O_1, then the five sentence structures are:

(i) S + V,
(ii) S + V + C,
(iii) S + V + O,
(iv) S + V + O_1 + O_2,
(v) S + V + O + C

These five structures normally correspond to five structures in Japanese:

(i) S WA V SURU,
(ii) S WA N DA or S WA A.

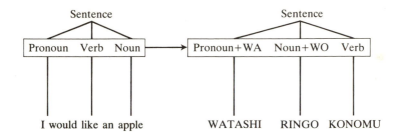

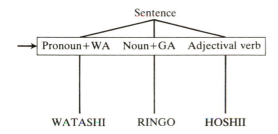

Fig. 4.1 The process of translation of a simple sentence.

The difference between these two expressions lies in the use of a noun (N) or an adjective/adjectival phrase (A).

(iii) S WA O WO V SURU or S WA O GA A DA.

Here, the two expressions are distinguished by whether the verb is V a verb or an adjectival.

(iv) S WA O_1 NI O_2 WO V SURU.
(v) S WA O GA C DA TO V SURU.

These are the most basic interlingual transformational rules for changing from English into Japanese. If, therefore, it is possible to determine which of these structures an English sentence has, it is theoretically possible to construct a corresponding Japanese sentence using these transformational principles. For example, the sentence:

I showed him a very interesting picture.

has the structure, S + V + O_1 + O_2. Provided we understand that 'a very interesting picture' is the direct object, O_2, the corresponding Japanese sentence has the following structure:

私は彼に大変面白い絵を見せた。
WATAKUSHI WA KARE NI TAIHEN OMOSHIROI
E WO MISETA.

I as for him to very interesting picture showed.

Most sentences in real situations are of course much more complex than these examples, but fundamentally, the central process in translation is the identification of which of these sentence structures a given verb is used in. For such recognition, it is first necessary to determine accurately which phrases are to be labeled S, O, C, etc.

Unfortunately, noun phrases can have varied and complex structures, including noun phrases which contain sentences embedded within them. The noun phrase 'a flower which is beautiful and radiant' contains such an 'embedded sentence' as a subordinate clause. As a consequence, a situation is frequently encountered where, in order to analyse a sentence, its noun phrases must be analysed, while in order to analyse the noun phrases, their sentence structures must be determined.

The grammatical rules of the structure of simple noun phrases in English can be written as follows:

Noun phrase = noun
            = article + noun
            = adjective + noun
            = adverb + adjective + noun
            = noun + preposition + noun
            = noun + relative pronoun + subordinate clause
            = ... etc.

Using these kinds of grammatical rules, it is possible to analyse the structure of sentences and the structure of phrases in similar ways, but there are some notable differences in the two kinds of analysis.

The easiest kind of example to understand is the analysis of sentences in which the word-order can be changed. For example, in the following Japanese sentences, the actual order of the words within the sentence can be rearranged at will:

私は昨日新幹線で東京へ行った。
WATASHI WA KINO SHINKANSEN DE TOKYO
E ITTA.

I as for yesterday Bullet-train on Tokyo to went.

昨日私は新幹線で東京へ行った。
KINO WATASHI WA SHINKANSEN DE TOKYO
E ITTA.

yesterday I as for Bullet-train on Tokyo to went.

私は昨日東京へ新幹線で行った。
WATASHI WA KINO TOKYO E SHINKANSEN DE
ITTA.

I as for yesterday Tokyo to Bullet-train on went.

With regard to the more difficult question whether these sentences express the same meaning, it can be said that the denotative contents of the three sentences are the same, but, as discussed in chapter 3, the attitude of the speaker and the emphasis within the sentences vary somewhat. Rather than get involved in a discussion of the translation of such subtleties at this point, let us consider only the translation of the denotative contents of these sentences. For this purpose, these three sentences have the same basic meaning and must therefore produce the same results when subjected to structural analysis.

In the analysis of Japanese sentences such as these, there is relatively little significance in the actual word-order of noun phrases. As a consequence, since it is necessary to have a method for performing morphological analysis for languages where word-order is unimportant, it is essential to understand that the analysis of phrases which include verbs is fundamentally different from the analysis of simple noun phrases. It has been found that case grammar is useful for the analysis of Japanese sentences. This will be discussed more fully in the following chapter.

The analysis of a sentence depends on using grammatical rules, such as those above, and deciding upon a structure which shows an overall coherence for all of the individual words that make up the sentence. This process is called syntactic analysis. When there are many rules of grammar, however, it becomes difficult to decide upon the order of application of these rules for defining the sentence structure. Since there is a great variety of possible sentences, the syntactic analysis proceeds by trial-and-error application of such rules until a balanced, coherent overall structure is obtained. In addition to those already mentioned, a variety of other techniques for increasing the efficiency of such analysis are under development, and this remains one of the most interesting areas of research, where significant improvement in the efficiency of machine translation systems should be obtainable.

By one means or another, (i) the structure of the sentence must be decided upon, (ii) the individual terms in the sentence must be replaced by the corresponding words in the target language, and (iii) based upon grammatical rules for transforming one language into another, the structure of the sentence must be changed into a structure which is appropriate for the target language. The internal structure of the sentence in the target language can thus be established. The reconstruction of the sentence in the target language from its internal structural representation can then be undertaken, a process known as sentence generation or synthesis.

When the internal structure has been found to be relatively straightforward, it is not difficult to achieve sentence generation. For example, when the sentence has a tree structure, as in Fig. 4.2, it is possible to perform the translation by starting at the word furthest to the left in the tree structure and proceeding to the right. The simplicity of this process is due to the fact that the tree has the same linear order as the the sentence itself.

The overall procedures involved in machine translation are illustrated in Fig. 4.3. The actual number of times which the system must cycle through analysis, transformation, and sentence generation depends upon the number of possible transformations. When a sentence undergoes syntactic analysis and is transformed into an internal structure, the various parts of the internal structure have as many possible substructures as the number of rules of grammar which have been applied to them. For each of these substructures, transformations must be done into the structure of the target language, and the transformed

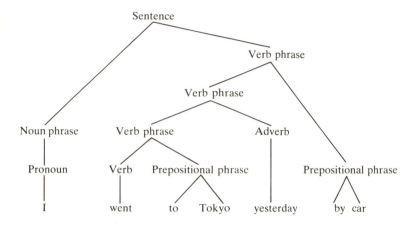

Fig. 4.2 The tree structure for 'I went to Tokyo yesterday by car.'

substructures must be organically fused once again to produce the internal structure in the target language. Once the internal structure is decided on, the final translation must be generated. In other words, by means of these repeated processes, and using a combination (finite) of grammatical rules, a virtually infinite number of linguistic expressions can be handled.

Problems in syntactic analysis

The fundamentals of translation are as I have described, but in practice this procedure includes many complex steps. Let us consider the process of translating from Japanese into another language, such as English. First, analysis of the Japanese must begin by decomposition of the given sentence into its word units. This is called morphemic analysis.* To determine where a word starts and ends, a series of adjacent characters must be extracted provisionally, and the presence or absence of that series in the word dictionary then investigated. If it is found to be present in the dictionary, the part of speech for that word is searched for,

---

*Unlike English and other Indo-European languages, the words in written Japanese are not broken into individual units by preceding and succeeding spaces, so that a preliminary stage of some complexity is required in Japanese.

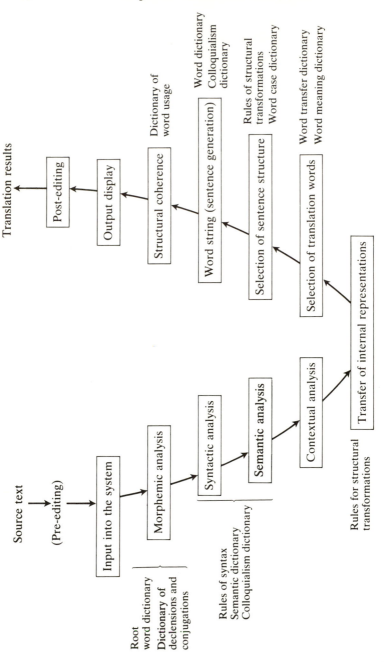

Fig. 4.3 The process of machine translation. Various dictionaries are used at each stage. Post-editing of the machine translation output is done by a human translator. In some cases the process can be simplified by pre-editing.

and then a check must be done to determine whether or not the conjunction of that word and the neighbouring words is grammatically permissible.

Only once the phrase or sentence is found to be grammatically coherent is it decided that a string of characters provisionally abstracted is a valid root as found in a dictionary. If these conditions are not satisfied, the following character is included for a renewed dictionary search. This troublesome process is undertaken repeatedly, and the continuous sequence of characters is eventually broken down into word units. At this point the parts of speech of these words and their inflection, declension, or conjugation in the sentence are known. Of course, when searching for various potential word units within a sentence, it is necessary to determine whether the ending of that character string is a suffix of some kind; if so, the root form of the word is used for the dictionary search.

Once the morphemic analysis has been completed, the analysis of the syntactic structure of the sentence itself begins. The basics of this stage of processing have already been discussed, but it should be noted that, in practice, several hundred rules of grammar are required for the analysis of the various kinds of sentence.

Research on ways in which to improve the efficiency of the application of such rules is an area of particular interest. It is known that the most difficult problem for sentence analysis is the resolution of ambiguities: that is, not in the application of rules of grammar, but when two or more different rules could be applied.

Ambiguous structures which frequently arise are of the following kind. The original expression may be 'A of B of C'. Here, two possible ways in which one noun phrase may refer to the others are evident: ((A of B) of C) and (A of (B of C)). In English, a similar problem arises with phrases such as 'the old man's hat'. Is it the hat or the man that is old? Referential ambiguities frequently arise and become more problematical as the number of elements increases. A of B of C of D can be interpreted as (((A of B) of C) of D), ((A of B) of (C of D), (A of ((B of C) of D))), ((A of (B of C)) of D), and so on. This type of problem is found in both English and Japanese.

Consider a related problem involving a prepositional phrase:

Verb + (direct object + prepositional phrase)

in which the prepositional phrase refers to the direct object; and

(Verb + direct object) + prepositional phrase

in which the prepositional phrase refers to the verb. Classic examples of these two cases are:

I bought a car with four doors.

I bought a car with four dollars.

And how will a machine translation system handle:

I saw a woman in the garden with a telescope.

Does the woman have the telescope or is it simply in the garden? A similar ambiguity arises when there are two possible interpretations of a phrase with the following structure: Adjective + noun and noun. One is ((Adjective + noun) and noun), and the other is (Adjective + (noun and noun)). For example:

peaceful atmosphere and crisis = (peaceful atmosphere) and crisis

rapid decision and execution = rapid (decision and execution)

In Japanese, as well, such examples are abundant:

電子　回路　の　電圧　と　電流
DENSHI KAIRO NO DEN'ATSU TO DENRYU

electronic circuit of voltage and current

which is translated as 'the voltage and the current of the electronic circuit'. A similar sentence structure in Japanese, however, might require a different grouping of the words:

電子　回路　の　雑音　と　全体　システム
DENSHI KAIRO NO ZATSUON TO ZENTAI
SHISUTEMU

electronic circuit of noise and entire system

which is translated as 'the electronic circuit's noise and the entire system...'). Consider this example, as well:

A, B, C 三種類の接着材についてのせん断強さ、剥離強さ、
および付着耐久度などの機械的性質

A, B, C SANSHURUI NO SETCHAKUZAI NI
TUITE NO SENDAN TSUYOSA, HAKURI
TSUYOSA, OYOBI FUCHAKU TAIKYUDO
NADO NO KIKAITEKI SEISHITSU

A, B, C three varieties of glue concerning shearing strength,
bonding strength, and durability, etc. of mechanical properties

which would be translated as 'the mechanical properties, such as
shearing strength, bonding strength and durability, of three
varieties of glue, A, B, and C'.

In this last example, if we consider the various combinations of
word-order and referential relationships simply at the level of the
parts of speech, already a large number of possibilities arise.

Many ambiguities also arise with regard to the interpretation of
individual words. In Japanese, the so-called adjuncts (such as
DE) can be used as auxiliary verbs or as particles which have
identical appearance:

会うのが目的でやって来た。
AU NO GA MOKUTEKI DE YATTE KITA.

meeting the fact of purpose with doing came

which is translated as 'I went with the purpose of meeting'. Here
DE is the conjunctive form of the auxiliary verb DA. In other
contexts, however, DE can function as a post-positional article:

わずかのお金でそれが買える。
WAZUKA NO OKANE DE SORE GA KAERU.

a little of money with that can buy

That can be bought for little money.

Here, the particle DE expresses the means by which buying will
occur.

There are also many cases in which the same article has different meanings. Here, NI expresses the locality of friends:

東京に友達がいる。

TOKYO NI TOMODACHI GA IRU

Tokyo in friends are

I have friends in Tokyo.

In a different example, however, NI indicates the recipient of the action:

三人におみやげをあげた。

SANNIN NI OMIYAGE WO AGETA.

three people to presents gave.

I gave presents to three people.

The question of how to eliminate these kinds of ambiguity has been a central theme since the beginnings of machine translation research. It is immediately obvious to anyone who considers the problem that, for a solution to be found, the meanings of words must be dealt with. But it is also true that semantics remains a massive problem for computers. Until the mid-1960s, various research efforts in machine translation were undertaken from the standpoint that the *analysis* of sentences (i.e., the most difficult step in the process of machine translation) might be accomplished working solely within the realm of syntax and without involving the troublesome problem of semantics. Unfortunately and despite tremendous progress in syntactic analysis, such a method was never discovered, and since the early 1970s there has been a gradual, if reluctant, recognition that semantic methods — not dissimilar to those which I attempted for sentence generation in 1963 — are unavoidable.

Semantic resolution of syntactic ambiguities

There are various ways of thinking about the understanding of the meaning of words. This will be discussed in detail in the following chapter, but here I will discuss the expression of the meaning of words by means of semantic primitives, which I and many others have employed in machine translation systems. It has been found that the technique of semantic primitives does not have to be developed to the maximum possible level of detail to be useful in determining the meanings of the words in a sentence. On the contrary, the chief purpose of the semantic primitives is simply to eliminate the various ambiguities which can arise in the interpretation of sentences.

As a consequence, the semantic primitives are constructed in the roughest and most incomplete manner that will still allow for this function. In other words, they constitute an entry into the world of meaning at the simplest level necessary for the successful analysis of sentences. This unsophisticated attitude has not been warmly welcomed by academics, but it is an inevitable development in the construction of practical systems which can analyse various kinds of sentence smoothly and consistently. Since thorough elucidation of semantic phenomena is still extremely difficult, the semantic primitives approach can be said to be, from an engineering point of view, the sounder way forward.

The fundamental idea behind this approach is that words with broadly similar meanings can be grouped together, and further subgrouping can be done on the basis of various selected semantic factors. There is in fact no need for the various subgroups to be mutually exclusive, and any single word might belong to a number of subgroups. The factors which are used for the subgrouping process are the essence of the semantic primitives, and words which are thought to have some degree of semantic commonality are grouped together. There is in fact little theoretical basis for the overall structure of the semantic primitives, and alternative systems will differ according to differences in the perspectives of researchers. Nonetheless, in many cases, groupings such as the following are used:

animals, humans, plants, inorganic matter, phenomena,
places, time, ... etc.

Figure 4.4 illustrates groupings of this kind where an attempt has been made to tie them together into a coherent system which covers essentially all varieties of objects and phenomena.

In an effective machine translation system which makes use of

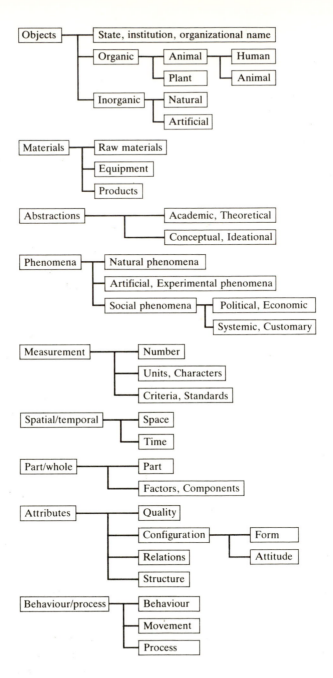

Fig. 4.4 An example of a table of semantic primitives.

semantic primitives, each word is assigned primitives, such as those shown in Fig. 4.4. For example, information must be given to the system such that it knows that 'Tokyo is a place', 'George is a man', etc. Thus, when analysing a series of noun phrases, those nouns which have related semantic primitives are more appropriately placed together than those with different meaning structures. It is worth noting, however, that a somewhat different semantic classification will often be required for a narrow technical field than for a dictionary of general terminology. This enables correct inferences to be drawn with regard to the structure of phrases, such as discussed earlier:

(the voltage and current) of an electronic circuit

the whole system and (the noise of the electronic circuit)

Similarly, because both 'car' and 'door' are products of human labour, they will probably be found in the same broad semantic group (for example, manufactured artefacts), whereas 'car' and 'dollar' are likely to have a more distant semantic relationship. Therefore, using meaning tables, correct inferences can also be drawn with regard to the structure of:

I bought (*a car with four doors*).

and:

I *bought* a car *with four dollars*.

The different meanings of the Japanese article NI can also be distinguished by making use of semantic primitives. Since 'Tokyo' is a place, the NI following 'Tokyo' is likely to indicate a place and 'Tokyo' is understood as a possible destination. The NI in SANNIN NI (to three people), on the other hand, is more likely to indicate a recipient of the action of the main verb 'to give', since SANNIN specifies the presence of people. This distinction between different uses of NI will be valuable when selecting the appropriate preposition for the English translation.

When trying to determine the structure of sentences using semantic primitives, it is not enough to know simply that two nouns are, for example, classified in the same group; it is also necessary to have information on the relationship between the two classes of words themselves. In other words, the nature of the relationship between the words must be made explicit in the tables, for example, higher or lower concepts, synonym or antonym, etc. A superordinate concept for the word 'children' is 'human being', and a superordinate concept for 'human being' is, in turn, 'animal'. Subordinate concepts for 'children' include 'infant', 'student', etc. A table of semantic primitives which includes the various interrelations among terms such as these is called a thesaurus, and it has been found that a thesaurus provides a powerful informational base for the mechanical translation of words. For example, since 'electronic circuit' and 'noise' will be stored in the thesaurus as related terms, and 'circuit' and 'system' will also be related, one of several possible interpretations is likely to be the strongest for:

((The noise of the electronic circuit) and the entire system).

It must be said, however, that because of the complex relations among words, the construction of a satisfactory thesaurus is not an easy matter.

In order to resolve the problems of the ambiguity of referential relationships, information concerning the likely relations among words must be stored in the computer in advance. For example, the word 'beautiful' is likely to be used to qualify various nouns denoting human beings, such as 'girl', but it is unlikely to be used to describe other nouns meaning human beings, such as 'kinsmen'. For this reason, it is necessary to place adjectives into various subgroups and then record the likely interrelations of the subgroups of adjectives and the subgroups of nouns.

The role of semantics in the analysis of sentence structure

Very few researchers in machine translation and linguistics show genuine confidence when it comes to describing precisely the semantic interrelations between word groups or even the semantic relations between words. Although it is generally easy enough to answer yes or no in response to a question whether a particular modifying relationship can exist between them, it is virtually impossible to record the relationship between each and every individual word that is to be stored in a dictionary.

Therefore, rather than work at the level of the individual word, it is unavoidable that the modifying relations be specified in terms of the relationships between two groups of words. The problem then arises that the specific modifying relationship will not always apply equally well to all of the words in the two groups, as noted above regarding 'beautiful girl' and 'beautiful kinsmen'. The result is that the semantic relationships which are specified in a thesaurus are not entirely accurate, but must be considered only approximate. Rather than have no semantic input at all, these semantic relations should be regarded as information which generally will suffice to produce good results in the process of analysis of sentence structure. In other words, this kind of semantic information must be used in full recognition of the fact that it is invariably somewhat inaccurate.

For the analysis of an actual sentence, relevant semantic information will often not be stored in the word dictionary, and cannot therefore be exploited in the analysis. The reason for there being such cases is that only the most likely semantic relations are included in the dictionary. When such information is not available, the machine translation proceeds with the ambiguities remaining, but this means that an extremely large number of ambiguities will often remain. Given a variety of possible translations, a reader is likely to be confused, so that machine translation systems are normally contructed to prefer those translations which have a theoretically higher probability of correctness in terms of sentence structure. For example, in the Japanese phrase:

美しい日本の私

UTSUKUSHII NIHON NO WATASHI

beautiful Japan of me

it is possible that either 'Japan' (NIHON) or 'me' (WATASHI) is qualified by 'beautiful' (UTSUKUSHII). In such cases, the preferential translation is based solely on the fact that 'beautiful' lies physically nearer to 'Japan' than to 'me' in the sentence. The result is that the sentence structure is interpreted as:

((UTSUKUSHII NIHON) NO WATASHI)

me from beautiful Japan

rather than:

(UTSUKUSHII (NIHON NO WATASHI))

beautiful me from Japan.

In any case, since semantic treatment is at all times somewhat uncertain, translations are not well produced when semantic information is given the highest priority in the analysis of sentences. Consequently, analysis based upon grammatical rules is normally given top priority, and only when an ambiguity in the syntactic analysis arises and it is unclear which of several possibilities should be chosen, is the semantic information utilized. When semantic information cannot then be used to resolve the ambiguity at this level, the sentence structure within the region of ambiguity which has the highest probability of being correct, is used.

5 Methods of analysis of sentence structure

Phrase structure grammar

A variety of methods have been devised for analysing sentences. The most fundamental technique is based upon the 'phrase structure' model advocated by Chomsky. This approach adopts the position that several neighbouring parts of speech together constitute a 'phrase', and indeed, the grammatical rules outlined in the previous chapter are based upon this conception of the phrase. For example, using formulae such as:

Noun phrase = adjective + noun
Noun phrase = noun + of + noun
Noun phrase = verb + noun phrase
Noun phrase = verb + noun phrase + noun phrase
Sentence = noun phrase + verb phrase

tree structures can be produced (Fig. 5.1). That is, the phrases to the left of the equality signs are defined as the roots within the tree structure, and the various elements to the right are called branches. By convention, the tree is read from left to right.

One other particularly important concept in Chomsky's grammar is the so-called transformational component. This concerns, for example, the relationship between sentences in English written in an active voice and those written in a passive voice. By means of transformational rules, the tree structure for an active verb can be transformed into the tree structure for a passive verb, or vice versa. Rules for such tranformations are typically of the following kind:

Subj (N1) + Verb + Obj (N2) ⇒ N2 + be + verb + verb (past participle) + by + Obj (N1)

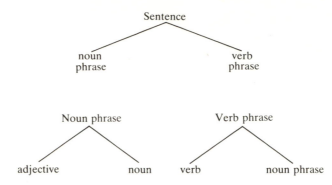

Fig. 5.1 The shape of a phrase structure tree.

When given a passive sentence, such rules can be used in reverse and the sentence analysed once it has been transformed into an active form.

In the midst of sentence analysis, ambiguities which emerge can be eliminated using semantic information, as was discussed in Chapter 4. The input of semantic parameters into the phrase structure grammar can be done in the following way. A formula such as:

Noun phrase = adjective (action) + noun (action)

can be applied to phrases such as 'rapid speed' or 'nimble movements', where both the adjective and the noun are typically used in describing physical actions. 'Rapid speed' and 'nimble movements' would therefore be seen, on the basis of this semantic information, as distinct phrases. When presented with the following Japanese phrase, however, a different conclusion would be drawn:

速い彼の駆け足

HAYAI KARE NO KAKEASHI

fast he of run

his fast run

In this case, 'fast he' is not considered a distinct phrase because there is no inherent semantic link between 'fast' and 'he'.

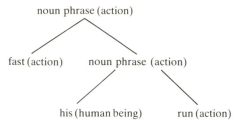

Fig. 5.2   Japanese and English tree structures for 'his fast run'.

Instead, 'he' is grouped with 'run' on semantic grounds, and 'fast' is seen as describing the 'run'.

If this were to be rewritten using tree structures, something like Fig. 5.2 would be obtained, semantic information being appended to each 'node' (branching point). Currently, many machine translation systems make use of grammatical rules which have semantic information appended on to the nodes of each phrase.

Transformations are also possible using a technique called 'predictive analysis'. This method is based upon *Grammar for the hearer* by Charles Hockett, one of the classic papers in linguistics, published in 1961.[10] Essentially, it is a hypothesis that when listening to speech, a human being will continually make predictions about the words and sentence structures that are to follow. For example, when a listener hears a sentence beginning with 'He', possible grammatical structures immediately come to mind: a verb might follow; or one form of the 'to be' verb or an auxiliary verb might follow; or the sentence might continue using one of the relative pronouns, who, whom, or whose.

This kind of situation can also be written as a set of grammatical rules. First, at the start of analysis, rules of the following kind can be written:

> Sentence = pronoun + verb phrase
> Sentence = pronoun + copula + complement
> Sentence = relative pronoun +
>   embedded sentence +
>   verb phrase

In other words, these rules indicate that if a sentence begins with a pronoun, it can be concluded with a verb phrase; or with a copula and complement; or with a relative pronoun, an embedded sentence, and a verb phrase; and so on. A verb phrase can also be written as rules, such as those which follow:

Verb phrase = verb + noun phrase
Verb phrase = verb + noun phrase + verb modifier

so that when a listener expects the appearance of a verb phrase, the actual appearance of a verb will mean that the sentence can be concluded with a noun phrase or with a noun phrase and verb modifier.

These kinds of grammatical rules were first used by Rhodes for the analysis of Russian text, and shown to produce extremely good results. Subsequently, Kuno wrote similar rules of grammar for the analysis of English text and demonstrated that even sentences which at first consideration appear to have a unique interpretation can in fact have several (see Chapter 2). In English, many nouns can also function as verbs, so that the number of possible structural ambiguities is quite large. The grammatical rules written for the predictive analysis method have capabilities essentially equivalent to those of Chomsky's phrase structure grammar, but they have the advantage of being applicable to the structure of sentences at any point in a sentence. That is, the rules can also be applied during the interpretation of a sentence, prior to hearing its completion.

One definite problem arises, however, in the handling of so-called garden path sentences, which require one to return to a point earlier in the sentence and reconsider the meaning of words located there. An example of a garden path sentence is:

The man sent to the house was a messenger.

If one reads only 'The man sent to the house...', 'sent' appears to be the main verb, but the moment the verb 'was' appears, it is evident that 'sent' is part of a subordinate clause which qualifies 'the man'. It is thought that not only machines but also human beings are forced to return to the start of the sentence for re-analysis whenever garden path sentences are encountered.

Whichever method is used, conventional phrase structure grammar or predictive analysis, problems arise with regard to the production of accurate translations and with regard to the num-

ber of rules which must be prepared in advance for the interpeta-
tion of the various allowed linguistic expressions.

Case grammar

In comparison with a language such as Russian, English has a
very small number of noun and pronoun declensions and verb
conjugations. In place of such inflections to the words themselves, the grammatical role of individual words is often determined by word-order within the sentence. In this sense, English
is a language which relies heavily on syntax. In Russian, due to
the nature of the numerous suffixes, the role of each word in the
sentence is determined almost regardless of where in the sentence
it is found. As a consequence, word-order does not carry as
much significance in Russian as in English. Because of the usage
of word adjuncts (suffix-like particles) in Japanese, the Japanese
language is more like Russian than English in this respect. As the
following examples show, however, the adjuncts following nouns
do not uniquely or unambiguously determine the role of the noun
in the sentence.

僕は魚が好きだ。

BOKU WA SAKANA GA SUKI DA

I as for fish liking is

I like fish.

それは僕が好きだ。

SORE WA BOKU GA SUKI DA

that as for I liking is

I like it.

Given that word-order in Japanese is fairly arbitrary, use of
phrase structure grammar is inconvenient for the analysis of
Japanese sentences, since it requires that individual rules must be
written for all possible permutations of sentence word-order.

   In light of such difficulties, a grammar based upon a completely
different way of thinking is normally used for the analysis of
Japanese. This method is called 'case grammar', as was briefly

mentioned in Chapter 3. In case grammar, verbs play a central role. Whenever a verb is encountered in a sentence, three essential questions are asked about it: (i) What is the subject (or actor) of that verb? (ii) What is the patient or object of the action of the verb? and (iii) What is the means by which that action is carried out? For example, the verb 'to open' is typically used in the following way:

A man opens the door with a key.

人が扉を鍵で開ける。
HITO GA TOBIRA WO KAGI DE AKERU.

The actor is the noun, 'man', the patient of the action is the noun denoting a manufactured artefact called a 'door', and the tool by means of which the door is opened is a 'key'. Thus, in any of the following sentences (and others like them) the relationships between the main verb 'open' and 'man', 'door', and 'key' do not change, only the final expression of the sentence is altered.

The man opened the door with a key.

The man's key opened the door.

The door was opened with a key by the man.

In the second sentence above, 'key' is in the normal position of the grammatical subject of the verb, whereas, in the third sentence, the grammatical subject is 'door'. These differences are, however, interpreted as merely differences in the surface form of the linguistic expressions, while the semantic interrelations among the words are identical in all three cases. The semantic relations can therefore be said to constitute the 'original structure' of the sentence which is in the speaker's mind before he actually speaks. The meaning can therefore be considered a deep structure underlying the final linguistic expression.

The relationships which a verb has with the neighbouring noun phrases will differ according to the meaning of the verb. A verb such as 'open' can have nouns which fill the roles of actor and object of the verb action, as well as those which indicate the means of the action. They are called respectively the agent case, object case, and instrument case. Case grammar is therefore essentially a technique of labelling each term in a sentence according to its relationship to the verb.

The first problem which is faced using such a technique is the question of how many different cases need to be represented. That is, unlike the small number of grammatical cases which are possible for nouns, the number of semantic case categories is not known with certainty. Fillmore, the first advocate of case grammar, hypothesized the existence of eight cases (Table 5.1), but the number of possible cases typically differs with the individual researcher using the technique. In our own work, we have distinguished a total of thirty-three different cases (Table 5.2).

The question of which cases are required by a given verb is answered by considering whether or not sentences can be constructed without the presence of each element. Those cases such as agent and object which cannot be excluded for a given verb are called 'essential cases', whereas cases which are not necessary for sentence construction are called 'inessential' or 'arbitrary'. For example, since the time and place of the action is unimportant for grammatical sentence construction with most verbs, cases related to the temporal dimension of the verb action are classified as inessential.

Whether terms are essential or inessential, the central process

Table 5.1 The cases used by Fillmore.

Agent (A)	the actor of the event
Counter-agent (C)	the force working against the event
Object (O)	the object which is moved or changed or whose position or presence is considered
Result (R)	the object which exists as a result of the action
Instrument (I)	the stimulus or direct physical cause provoking the event
Source (S)	the location at which movement has originated
Goal (G)	the location to which movement has been made or results have moved
Experience (E)	the object which, as a result of the action, has experienced something

Table 5.2   The 33 cases used in the analysis of Japanese

Label	Example
(1) subject	*The boy* walked home.
(2) object	She found *the book*.
(3) recipient	gave to her.
(4) origin	received from him.
(5) partner	to consult with . . .
(6) opponent	to protect from . . .
(7) time	In 1980, . . .
(8) time-from	From May of last year, . . .
(9) time-to	Until next year, . . .
(10) duration	Over a period of five minutes, . . .
(11) space	. . . is located at . . .
(12) space-from	to return from . . .
(13) space-to	to send to . . .
(14) space-through	to pass through . . .
(15) source	to translate from Japanese
(16) goal	to translate into English
(17) attribute	to be rich in . . .
(18) cause	to be due to . . .
(19) tool	. . . with a hammer . . .
(20) material	to be made out of . . .
(21) component	to consist of . . .
(22) manner	at a rate of . . .
(23) condition	to determine under the conditions of . . .
(24) purpose	adapted to . . .
(25) role	to use as . . .
(26) content	to be seen as . . .
(27) range	with regard to . . .
(28) topic	as for the topic of . . .
(29) viewpoint	from the perspective of . . .
(30) comparison	better than . . .
(31) accompaniment	together with . . .
(32) degree	an increase of 5%
(33) predication	. . . is . . .

required for analysis of a sentence is determination of the various terms located in the vicinity of a given verb. It should now be evident that the analytic process depends more on the semantic relationships among the words, their place in a table of semantic primitives, and the nature of any word suffixes than on their order within the sentence. The representation of the essential cases for each verb and the kinds of nouns which those cases take, based upon a semantic table, is called the 'case frame' of that verb. In the representation of verbs in the computer dictionary entry, there is a description of the case frame unique to each verb.

The actual analysis of a given sentence begins with identification of the main verb. The next step — based upon semantic information, adjuncts, and information obtained from the declensions of nouns and conjugations of verbs — is to determine which case frame of the given verb each of the noun phrases in the sentence belongs to (that is, the 'case slot').

Similar ideas can be applied not only to verbs, but also to adjectives and, in Japanese, to nouns which are constructed by the addition of -SA. For example, TSUYOSA (strength) is a -SA noun derived from the adjective TSUYOI (strong). The -SA nouns in Japanese are analogous to the '-ness' nouns in English, but the -SA nouns are employed far more frequently. For example, the adjective 'beautiful' can be transformed into a noun by appending 'ness': 'beautifulness'. For a Japanese adjective such as UTSUKUSHII (beautiful), the variety of nouns which are its likely referents are represented in a semantic table. When transformed into a noun (UTSUKUSHISA), the same case frame which is found when the word is used as an adjective would apply.

Often -SA nouns are usefully interpreted using the genitive particle NO, indicating possession:

noun 1 の_____さ

noun 1 + NO + -SA noun

The ___ -ness of the noun 1

Another common construction using NO is:

コンピュータの理解

Computer NO RIKAI

computer of understanding

since the agent of 'understanding' is likely to be a human being, the computer is given an object case, and the phrase is interpreted as 'the act of understanding computers', rather than 'the act of a computer understanding [something else]'. This example clearly indicates not only the possible ambiguity in assigning case, but also the fact that the evolution of language can require changes in assignments as a language changes.

Recently, however, due to developments in artificial intelligence and efforts to make computers more intelligent, the use of 'computer understanding' to mean 'the act of a computer understanding something else' has greatly. increased. Indeed, the English language phrase 'computer understanding' in the latter sense has become fashionable and is perhaps already the preferred interpretation. In this and similar cases, 'computer' must be handled as a quasi-human agent and be given the 'agent' case. In summary, case grammar is a grammar which accomplishes a (limited) fusion of semantics and sentence structure.

Procedures in sentence analysis

It is extremely important to consider the ways in which several hundred grammatical rules are to be used in the analysis of sentences. For the purposes of increasing the speed of analysis, avoiding the output of inessential analytic results, or finding out which grammatical rule is at fault when an analysis is not correct, it is convenient to group the grammatical rules according to the structure of the language being dealt with, and decide upon the sequence of application of rules.

Unfortunately, there is not necessarily a definite order to the analysis of sentences. This is because rules can be written to fit the structure of any sentence and there is no implicit hierarchy among all possible rules. From the general nature of the way in which the sentence is constructed, however, there is often a sequence of analysis which is most natural. Figure 5.3 outlines one example of the sequential order used in the analysis of Japanese sentences.

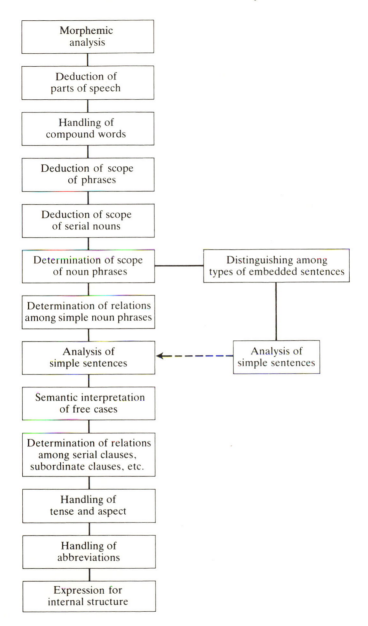

Fig. 5.3 The procedure used in analysis of Japanese text.

First, the given Japanese sentence is broken into its word units by the process called morphemic analysis. The parts of speech are then assigned to the individual words, and the inflection of the words is determined. The next stage is to decide upon the appropriate parts of speech for all words in the sentence which can be used in various capacities. In the early days of machine translation, determination of the parts of speech for each word was undertaken only at a late stage, once analysis of the entire sentence was completed, but that method was found to involve a huge number of fruitless and unnecessary steps. Consequently, before thorough structural analysis of a sentence is performed the parts of speech of the words which can be decided upon are specified and the characteristic features of the overall sentence are determined from its surface structure.

The most difficult aspect of such analysis when dealing with Japanese concerns the adjuncts. For example, a distinction must be made between the imperfect form of the auxiliary verb DA, which becomes DE, and the particle DE. If the correct distinction can be made, an accurate translation becomes possible (see Chapter 4).

The examples which follow were translated by our machine system, having successfully made such distinctions.

この回路はわずかの修正で、他の計算機にも接続できる。
KONO KAIRO WA WAZUKA NO SHUSEI DE,
HOKA NO KEISANKI NI MO SETSUZOKU
DEKIRU

this circuits as for slight of correction by, other of computers
to also connect can be done

This circuit can also be connected with other computers
by slight correction.

この方法はR・ベルマンの理論から道いたもので、
解析幾何理論の応用である。
KONO HOHO WA R. Bellman NO
RIRON KARA MIICHIITA MONO DE,
KAISEKI KIKA RIRON NO OYO DE ARU

this method as for R. Bellman of theory from derived thing is, analytical geometry theories of application is

This method is one derived from theories of R. Bellman, and this method is an application of analytical geometry theories.

Another problem, which is particularly difficult when Japanese sentences have been constructed using the conjunctive form of a verb, concerns the deduction of the scope of the phrases which must be translated in connection with one another. For example, consider the translation by machine of the following sentence:

計算機で基板のパターン設計を行い、
製品までの格固定を生業する
システムを検討。

KEISANKI DE KIBAN NO pattern SEKKEI
WO OKONAI,

SEIHIN MADE NO KAKUKOTEI WO SEIGYO
SURU system WO KENTO

computer by substrates pattern design carry out, product as far as
arrangement control system to study

The English which emerges from the machine translation system is:

The systems are studied which carry out the pattern design
of substrates by computers and control each process
from arrangement of parts up to products.

While a more flowing style might be desired, the translation was successful in determining that the two consecutive phrases (ending in OKONAI and SEIGYO SURU) both qualify system ('systems which carry out...' and 'systems which control...'). There was clearly the possibility of breaking the sentence differently into '...OKONAI' (carry out design) and 'system WO KENTO' (study the system). In translation, this would have been incorrectly rendered in English as 'Performing pattern design of substrates by computer and studying systems which control each process for arrangement of parts up to products.'

It should be noted, however, that it is probably incorrect to view the leading phrase, KEISANKI DE (by computer), as referring only to the SEKKEI WO OKONAI (carrying out the pattern design) phrase. On the contrary, the 'by computer' phrase refers to both of the phrases which followed ('carry out...' and 'control...'). In this respect, the analysis by the machine translation system was incorrect. In order to obtain a correct analysis of this point, however, decisions would need to be made about the meaning of the sentence (specifically, about what computers are capable of doing), and currently there is no general method for dealing with such sophisticated semantic information.

Since at least two verbs (predicates) are involved when the conjunctive verb form is used, decisions must be made concerning whether or not the verbs can be arranged serially with the same subject or object. Such decisions are made on the basis of the commonality of the semantic primitives of the verbs.

Here, an important distinction in determining the commonality of the verbs can be made according to the 'intentionality' or 'non-intentionality' of the verb. This refers to whether or not the agent's 'will' is implicitly contained in the verb itself. For example, verbs such as 'become' or 'to be changed into' are seen as verbs which do not include the intentions of the actor of the verb, whereas 'to do' or 'to change' do include such intentions.

We have already briefly discussed the problems encountered in the analysis of sequences of noun phrases in Chapter 4, but it deserves emphasis that the determination of the scope of complex noun phrases and the correct analysis of such structures is a formidable problem. Particularly in technical-scientific papers and their abstracts, complex linguistic expressions of a peculiar nature are frequently found. For example, even for a normal Japanese reader, a correct understanding of a sentence such as the following is difficult in a single reading:

国際電気委員会、アメリカ国家規格、衛生規則による
人体に対する許容接触電圧および電流
の規定、および主要な防御系を述べる。

KOKUSAI DENKI IINKAI, America KOKKA
KIKAKU, EISEI KISOKU NI YORU JINTAI NI
TAI SURU KYOYO SESSHOKU DEN'ATSU
OYOBI DENRYU NO KITEI, OYOBI SHUYO
NA BOGYOKEI WO NOBERU

international electronics society members, American national
standards, health rules based upon allowed human limits
for voltage and current regulations, and major protective
systems to discuss

The output from our machine translation system was:

The regulation for permissible contact voltages and currents by
International Electrotechnical Commission, USA national
regulation, health regulation to human bodies and major
protection systems are described.

It should be evident that it is somewhat difficult even for a
human translator to decide upon the scope of the individual noun
phrases and where they refer within a sentence such as that
shown above. One is tempted to go back and ask the original
author just what his true meaning was! More realistically, let us
simply say that input of such complex sentences into a machine
translation system should be avoided whenever possible.

Embedded sentences

Another difficult problem in the analysis of sentences is the
handling of embedded sentences. For example, 'I saw the letter
which he received.':

私は彼が受け取った手紙を見た。
WATASHI WA KARE GA UKETOTTA TEGAMI
WO MITA

I as for he received letter have seen

Within the sentence 'I have seen the letter', there is the de-
pendent clause 'which he received'. It can be considered as an
embedded sentence if it is rephrased as 'The letter he received is
the letter which I have seen.'
    Normally, in embedded sentences in Japanese a noun phrase
will be omitted, due to the fact that the noun phrase is placed at
the end of the embedded sentence. In other words, the above
sentence can be rewritten as:

WATASHI WA KARE GA △ UKETOTTA
TEGAMI WO MITA.

in which the triangle signifies the location of the direct object of the embedded sentence ('TEGAMI WO'), which has been omitted. Similarly, the following sentences have noun phrases, which have been omitted:

彼が手紙を受け取った日は雨だった。
KARE GA TEGAMI WO UKETOTTA HI WA
AME DATTA

He letter received day as for rain was

which can be rephrased as:

彼が△手紙を受け取った日は雨だった。
KARE GA △ TEGAMI WO UKETOTTA HI WA
AME DATTA

He △ letter received day as for rain was

where the triangle corresponds to 'the day on which'.

手紙を受け取った彼はよろこんだ。
TEGAMI WO UKETOTTA KARE WA
YOROKONDA

letter received him as for overjoyed

which can be rephrased as:

△手紙を受け取った彼はよろこんだ。
△ TEGAMI WO UKETOTTA KARE WA
YOROKONDA

△ letter received he as for overjoyed

where the triangle corresponds to 'he'.

For sentence analysis, there is a need for reconstruction of the complete sentence without omissions, even if omissions would be the normal condition in ordinary text. Since the adjuncts of

Japanese (WA, NI, GA, WO, ...), which would normally follow the noun in the embedded sentence, indicate the cases of nouns relative to the verb (which is placed still further along in the sentence), the essential information indicating the case of the omitted noun in the embedded sentence is also missing.

As a means for inferring that information, it is necessary to look up the case frame of the verb found in the embedded sentence in the dictionary, to determine what cases can precede the given verb. A check is then made to determine if it is possible that the following noun phrase can be qualified in that way. For example, if we have a missing direct object case with, for example, the verb UKETORU, in order to determine whether or not the following noun phrase, TEGAMI, can be in the patient (direct object) case, it is ascertained whether or not the noun TEGAMI has semantic primitives such as those required for a direct object slot (TEGAMI WO UKETORU).

In the case of TEGAMI WO UKETOTTA KARE WA YOROKONDA, it is the agent case for the verb UKETORU which is missing, and analysis must be made to determine whether or not the following pronoun KARE can be used as the subject of the verb UKETORU. Although far from colloquial speech in Japanese, the question must be asked whether or not it would be valid to say KARE GA TEGAMI WO UKETOTTA KARE WA YOROKONDA. The English equivalent would be 'He who received the letter is he who was overjoyed.'

Similar checks must be made when other case slots are unfilled.

We have now briefly discussed the noun cases in sentences in which the noun phrase is immediately preceded by a verb which modifies it, but there is a slightly different form of the embedded sentence which must also be dealt with. For example:

処理速度が速い計算機を使う。
SHORISOKUDO GA HAYAI KEISANKI
WO TSUKAU

processing speed is fast computer to use

He uses a computer with a fast processing speed.

In this example, there are no empty slots for the Japanese predicate meaning 'fast' (HAYAI). The qualified noun,

KEISANKI (computer), is itself qualifying SHORI SOKUDO (processing speed), so that the sentence as a whole could be rewritten as:

計算機の処理速度が速い、そういう計算機を使う。
KEISANKI NO SHORI SOKUDO GA HAYAI,
SO IU KEISANKI WO TSUKAU

computer of processing speed is fast, that kind of computer to use

He uses that kind of computer which is a fast processing speed computer.

In other words, that which has been omitted in the original sentence is a noun phrase indicating 'computer' preceding the entire sentence:

△の処理速度が速い計算機を使う。
△ NO SHORI SOKUDO GA HAYAI KEISANKI
WO TSUKAU

△ of processing speed is fast computer to use

As might be inferred from the obliqueness of such examples, in many cases of this kind it is quite a difficult matter to identify such relationships correctly and then reconstruct the embedded sentences. A conditional adjective or adjectival phrase often becomes the predicate of this form of embedded sentence, and the immediately preceding noun expresses possession or is a noun specifying an action. In this example, reconstructing the complete embedded sentence would give a form such as:

The [part, belonging to or action] of [the referent noun phrase] is [a conditional predicate].

In an example such as:

窓の明るい部屋
MADO NO AKARUI HEYA

window of is bright room

A room with a bright window

reconstruction would give:

部屋の窓が明るい

HEYA NO MADO GA AKARUI

room of window is bright

The room's window is bright.

A third kind of embedded sentence is relatively clear cut. The referent is a special noun indicating cause, purpose, circumstances, nature, conditions, methods, necessity, etc. Embedded sentences of the following form are found:

1. 地すべりが起った原因

   TSUCHI SUBERI GA OKOTTA GEN'IN

   landslide occurred cause

   The cause of the landslide

2. アメリカへ行く目的

   America E IKU MOKUTEKI

   America to going purpose

   The purpose in going to America

3. 勉強をやりはじめた動機

   BENKYO WO YARI HAJIMETA DOKI

   study do start to motivation

   The motivation for starting studies

In each of these cases, the referent is reconstructed in the embedded sentence using the conjunctival form of the verb, and takes the case slot appropriate for 'causes', 'purposes', etc.

There are several other unusual constructions for embedded sentences which would be extremely difficult to handle in mechanical ways. Fortunately, the possibilities for their use in scientific and technical documents are few.

The procedures employed for analysing Japanese language sentences have been discussed above and are summarized in Fig. 5.3. The flow of the analysis shown indicates the steps required for analysing a complete sentence. Since almost identical steps are required in the analysis of embedded sentences, the analysis of sentences containing embedded sentences implies that all of the steps in Fig. 5.3 will be employed, and then others. Specifically, the entire sequence of processing steps is required for the analysis of each embedded sentence, meaning that a backtracking or recursive procedure is needed. This is similar to using a function which is defined in terms of itself, and is a familiar trick in computer programming. By means of such functions in software systems for machine translation, sentences which appear at the outset to be extremely complex can often be analysed without difficulty.

Using case grammar, the necessary cases corresponding to the verbs in a given sentence can be decided upon, and the remaining noun phrases are left simply as 'open' cases. The correct cases for all open cases are eventually determined by the nature of the nouns and their associated words and phrases.

When analysing Japanese sentences, the next step is to process noun phrases containing WA, the peculiarly Japanese particle indicating the main thematic topic of the sentence (but not always the grammatical subject). Noun phrases followed by WA often have a related predicate located at a distant site — often at the end of the sentence, although there are exceptions to this rule. When there are two noun phrases which are connected by verbs in the conjunctival form and taking WA, a decision must be made on whether the noun phrase followed by WA refers to both predicates or only to the first one. In technical-scientific papers, there is a special kind of usage in which the predicate comes at the close of the sentence, and the author may omit the noun which normally would have appeared as the subject of the sentence. Cases in which inferences of this kind must be made are those where there is a noun phrase followed by WA, which does not refer to the final predicate but in fact refers only to the preceding predicate. For example, consider the following sentence, typical of those found in technical papers:

荷電粒子の運動は回析放射バーストをともなうことを
確認した。

KADENRYUSHI NO UNDO WA KAISEKI
HOSHA burst WO TOMONAU KOTO WO
KAKUNIN SHITA

charged particle of movement as for radiation burst accompanies
the fact that identification done

[We] have found that the movement of a charged particle
is accompanied by a radiation burst.

Here, those who did the 'identifying' were the (unmentioned)
authors, and the phrase KADENRYUSHI NO UNDO WA
refers to the verb TOMONAU (accompanies), rather than the
verb KAKUNIN SURU (identifies). It is even less easy to determine the appropriate verb for the WA phrase in the following
example:

研究者は子供をともなっていることを確認した。
KENKYUSHA WA KODOMO WO
TOMONATTEIRU KOTO WO
KAKUNIN SHITA.

researchers as for children accompany the fact that
identification done

Here, the WA phrase (KENKYUSHA WA) could refer to
either verb, TOMONATTEIRU or KAKUNIN SURU. In other
words, the Japanese phrase could be translated as either 'We
have found that the researchers accompanied the children' or
'The researchers found that the children were accompanied'.

The portion of Fig. 5.3 which illustrates the treatment of verb
tense and aspect entails quite complex manipulations. Tense refers to the facts indicating the temporal dimension of sentences,
and includes present, past, future, perfect, and imperfect conditions. Normally, tense is determined by the conjugational
changes in the predicate and the various terms (adverbs) related
to the verb. The temporal dimension and tense must always be
made to correspond accurately in the two languages. When there
is only one verb in a sentence, correspondence can normally be
obtained in a straightforward manner and a translation precisely
equivalent to the source language can be achieved.

It is far more difficult, however, to achieve translation when
complex embedded sentences are present. It is then essential not
merely to identify the tenses of the verbs in the source language,
but also to identify correctly the temporal relations between the
main verb and those found in embedded sentences.

Fortunately, due to recent linguistic work, the temporal correspondences between Japanese and English have been greatly clarified, and such theory has been sufficiently explicit to implement in computer systems. Moreover, since complex combinations of tenses are not commonly used in technical and scientific writing, it has been possible to simplify such linguistic rules to a minimum when processing for verb tense.

Finally, with regard to aspect, various phrases are used primarily to express the speaker's attitude. These include Japanese phrases such as:

SHITAI   want to...

NI CHIGAI NAI   it is certainly the case that...

SHITE MO YOI   it is all right if...

and English phrases such as:

must   SHINAKEREBA NARANAI

should   ...BEKI

want to   SHITAI

seem to   RASHII

In combination with the possibilities for tense, the correspondence of tense and aspect in main and embedded sentences between two different languages becomes quite complex, but again recent developments in linguistics have greatly clarified most problems in this realm.

6 The selection of words for translation and sentence generation

Intermediary expressions

Analysed by the steps outlined in the previous chapter, a sentence, originally merely a string of characters, can be transformed into a complex tree structure. This tree structure is often referred to as the 'internal structure' or 'deep structure' of the sentence (but not in the Chomskian sense). When analysis is performed using case grammar, the main branches of the tree structure indicate the relationships among the cases of the nouns in the sentence. A simple example is shown in Fig. 6.1.

An example of a sentence containing an embedded sentence within it is:

I saw the letter he received.

which has the tree structure shown in Fig. 6.2.

The results of analysis of the sentence shown in Fig. 6.1, this time using Chomsky's phrase structure grammar, can be expressed as in Fig. 6.3. At each of the nodes of the tree a grammatical term is used, and the sequential order of the branches in the tree has a particularly important meaning. In Fig. 6.3 two separate noun phrases appear, but from their relative positions it can be concluded that the first noun phrase is the subject and the second is the direct object of the main verb.

A comparison of the analytical results of phrase structure grammar and case grammar reveals that the left-right position of phrases in the trees of phrase structure grammar is important, whereas the order of the branches in case grammar is not as significant. Instead, it is the information found at each branch of the tree that is important in case grammar. It should be noted,

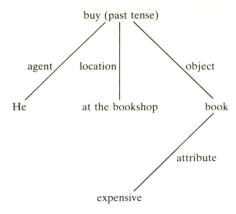

Fig. 6.1 The case grammar tree structure of 'He bought an expensive book at the bookshop.'

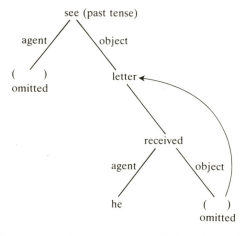

Fig. 6.2 The tree structure of an embedded sentence, 'I saw the letter he received.'

however, that even when case grammar is used for analysis, the order of the words in the original sentence is normally maintained in the relative positions of the branches of the tree. The results of case grammar analysis are therefore superior in so far as they contain more information.

Since the case relations of the internal structure of a sentence

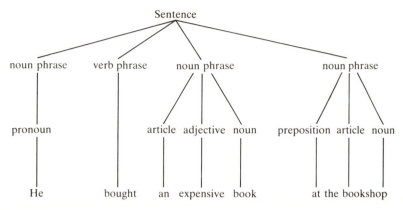

Fig. 6.3 The tree structure using Chomsky's phrase structure grammar of 'He bought an expensive book at the bookshop'.

can also be seen as expressing the fundamental meaning of the sentence, there are some researchers who emphasize that the *understanding* of a sentence is identical to determining the case structure of the sentence. In this sense, the case grammar representation of the internal structure can be regarded as a pivot or intermediate expression lying between the two natural languages for which translation is being performed. It is for this reason that case grammar is now widely used in machine translation system.

The representation of the internal structure of a sentence which has been analysed in terms of its case relations is found to be quite universal. Assuming that the term at each node of the tree represents a concept in the given language, it is possible to interpret the representation of the internal structure as expressing the interrelationships of the concepts in the sentence. These conceptual relations represent abstract universals which are quite distinct from the languages in which they are expressed. If these relations are identical no matter which languages are involved, the case grammar representation can function as an intermediary lying between any two languages. For this reason, it is frequently referred to as a 'pivot' language. The idea that the internal structure resulting from sentence analysis has the same structure regardless of which language it has been taken from means, in effect, that the representation of identical meanings will have identical internal structures. If this were strictly the case, then multilingual translation would be a comparatively easy matter. That is, if a sentence in a certain language could be analysed and an internal structure found, sentences with the same meaning could subsequently be synthesized in any other language. Moreover, the generation of such sentences would be completely unrelated to the idiosyncratic structure of the sentence in the original language, but would have the advantage of conveying only the information of the original internal structure.

A variety of research projects were undertaken in the 1960s on the basis of the supposition that this kind of ideal pivot language exists. Undoubtedly, one of the most significant influences pushing in this direction was Chomsky's hypothesis that, because of the universally similar structures of the human brain, processing of language must be fundamentally the same among all normal people capable of language. As a candidate for such an intermediary language, Chomsky advocated the 'deep structure' of language (the actual configuration of which has been reworked several times), while others have developed case grammar representations, and still others have worked on theories based upon symbolic logic.

It was found, however, that linguistic expressions show differences not only because of the idiosyncracies of the individual languages themselves, but also because of social and cultural conditions. As a result, even when translation into expressions with the same meaning is accomplished, there can remain important differences. As will be explained in greater detail in the section on structural transformations, the analysis of sentences with the same contents in two different languages will not always result in the same internal representation. When translation is attempted not simply of the fundamental semantic contents of the sentence, but also of the connotations which express the speaker's attitude and the social circumstances of the sentence, then the determination of a neutral intermediary linguistic expression becomes particularly difficult, precisely to the degree that such higher level considerations are included.

In this sense, it should be apparent that the idea of a pivot language — the expressions for which are chosen in light of semantic analysis, and from which translation into various languages can be done — is well beyond the level of current machine translation techniques. Even systems which are designed to follow such principles operate at a much lower, greatly simplified level. The pivot languages in current machine translation systems can extract only the simplest factual relationships among the words in a sentence, and all information of higher levels and subtle nuance is simply lost.

For these reasons, the possibility of a universal pivot language must be considered unlikely, and its pursuit has in fact been largely abandoned. In its place has appeared something referred to as the transfer technique, which involves translation of the internal structure of one language directly into the internal structure of another language, from which sentence generation proceeds. The process of changing from one internal structure to another will differ radically depending upon the source and target languages, and requires $n \times (n-1)$ internal-structure translation routines for translation among 'n' different languages.

What this means is that, for example, in the European Community, which had 9 recognized languages as of 1986, 72 different procedures for translating from one internal structure to another are required. Needless to say, preparation of those procedures alone is a massive undertaking, and efforts are therefore being made to reduce these transfer procedures to their simplest possible forms. In other words, as much detailed information as possible is abstracted, the analytic results are presented in as neutral a form as possible, and the transfer procedure from one language to another is reduced to the most rudimentary form.

This technique has been found to be feasible for the various languages of Europe, but it is totally unsuitable for two languages such as English and Japanese, which use completely different sentence structures. Rather than simplify the transfer procedure for these two languages, it is necessary to make it as detailed as possible. Only when the transfer procedure between the internal structures of Japanese and English is greatly elaborated is it possible to produce good quality translations in which a sentence structure very different from the original is produced in the target language.

Selection of words for translation

The transfer process for the internal structure of sentences in machine translation consists of two procedures: selection of

words among various possible translations, and structural transformation. The selection of words is essentially the problem of
deciding the term which corresponds most accurately to the term
in the original sentence. As a rule, there are several possible
translations of a given word. One familiar example is the English
word 'spring', which can mean: (i) a season of the year, (ii) a
coiled wire object, (iii) a natural source of water, and (iv) to leap
into the air. Thus the first problem in the translation of 'spring' is
to decide which kind of spring is involved. Selection of the
correct translation can often be made solely on the basis of the
field within which the translation is made. If we are involved in
translation of text in the field of mechanical engineering, 'spring'
is likely to mean a coiled wire object. In general, the narrower
the specialized field of the original text, the more likely is it that
individual words will have unique translations.

In mathematics, 'field' has one meaning, normally translated as
TAI in Japanese, whereas in physics or electromagnetics it has a
different meaning, normally translated as BA or KAI. It is a far
more difficult matter, however, to choose the correct translation
of words in non-specialist text, for example, in the sentences
found in magazines and newspapers written for a general read-
ership. 'Field' might then be translated in the sense of a field for
growing crops (HATAKE), or field in the sense of an open space
or plain (NOHARA), or a sports field (KYOUGIBA), or field in
the sense of a sphere of activity or specialization (BUN'YA), or
field in the sense of a specific scene (GENBA), or field in the
sense of the range of vision (SHIYA).

It is thus essential to make distinctions among such possibilities
on the basis of first principles, and to have a machine translation
system which is capable of automatically deciding which usage of
the word is being employed in any given sentence. This is no easy
task.

Deciding on the correct translation for verbs is an even more
difficult task than that for nouns. It is known that many subtle
factors can influence the choice of the appropriate verb, and such
complex processing is still beyond the capacities of machine
translation systems. In general, the selection of the verb is made
on the basis of the meanings of the subject for the verb and its
direct and indirect objects. For example, in translating from the
Japanese:

X が生じる

X GA SHOJIRU

X becomes

if X is a physical object or an idea, then the verb can be translated as 'to form X'. Therefore:

氷が生じる

KORI GA SHOJIRU

ice forms

is translated as 'ice forms → ice is formed' and

アイデイアが生じる

idea GA SHOJIRU

idea forms

is translated as 'an idea forms → an idea is formed'. If, however, we are dealing with social events, then the correct translation is likely to be 'X occurs'. For example:

事故が生じた

JIKO GA SHOJITA

accident occurred

would normally be translated as 'an accident occurred', whereas if the causal agent is known, a similar structure is seen in the Japanese, but a different translation into English:

X が Y を生じる

X GA Y WO SHOJIRU

X Y causes

Depending upon the nature of Y, the correct translation could be: X causes Y (behaviour), X forms Y (material transformation), or X produces Y (in most other cases).

In the case of many common verbs, such as 'do' and 'become'

Table 6.1 Examples of the translation of NARU into English.

Japanese expression	Meaning of b	English expression
a GA b TO NARU	means, tool	a provides b
	numerical value, unit	a reaches b
	theory, method, concept	a turns into b
	degree, level	a becomes b

(and their Japanese equivalents, SURU and NARU), there are an exceedingly large number of possible translations, and a thorough search for possible translations must be carried out. A small number of the possibilities for NARU are shown in Table 6.1. Unfortunately, selection of the appropriate translation is extremely difficult, and there are no general rules which can be used with confidence. As a consequence, the criteria required for making correct translations must be worked out individually for all such words. This is a major task linguistically, but it is a simple fact that high-quality translations cannot be achieved if the precise nature of individual nouns and verbs is not described on the basis of such criteria.

It should be noted that such information on individual words is not required solely for the purposes of machine translations, but is frequently required by people in other kinds of linguistic work. For example, students of foreign languages require a description which is as objective and general as possible of how words are used in the foreign language.

For these various reasons, the construction of dictionaries with detailed contents must be undertaken (and will take a long time). The single most important factor determining the usefulness of a machine language system is the quality, depth, and richness of the data in such a dictionary.

Structural transfers

When translation is to be performed between languages with extensive linguistic similarities, there is little need to change the internal structure of the sentence in the source language into the internal structure of the target language. For example, translation between French and English or between French and Italian will often reveal identical internal structures for two sentences of the same content. However, the internal structures for two languages such as English and Japanese, which have completely different historical and linguistic roots, will often prove to be very different. Consider the following example of English and Japanese sentences with the same meanings:

Many students made the same mistake.

同じ間違いをした学生が多かった。
ONAJI MACHIGAI WO SHITA GAKUSEI
GA OKATTA.

same mistake made students were many

Structural analysis produces completely different internal representations. This can be seen if we consider the literal translation (plus minor changes) of the Japanese into English or vice versa. The literal translation of the above English gives:

多くの学生が同じ間違いをした。
OKU NO GAKUSEI GA ONAJI MACHIGAI
WO SHITA

many of students same mistake made

And the literal translation of the Japanese produces an English equivalent as follows:

There were many students who made the same mistake.

Clearly, although the meaning is the same, the internal structures are different. What this implies is that if such literal translations will suffice, as in the above examples, then computer translations can be accomplished relatively easily. When the two languages involved are linguistically similar, such literal translations are likely to be heard as natural expressions in the target language, as well as in the source language. In the case of English and Japanese, however, such literal translation often will not suffice. On the other hand, since the Japanese people have been exposed to the sentence structures of literally translated English-language

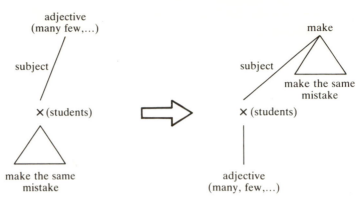

Fig. 6.4   An example of a general transformational rule.

works for many decades, there is perhaps little resistance among Japanese to hearing Japanese-language sentences with somewhat unusual sentence structures. It is a different matter to have a native English speaker read the English sentences of a text which has been literally translated from the Japanese! Which is to say that with languages with completely different linguistic constructions it is necessary to make appropriate transformations from the internal structure of one language into that of another.

If we consider the transformation of the above example from Japanese into English, it is necessary to make use of transformational rules such as the one shown in Fig. 6.4. Since this transformation is a very general one when translating from Japanese to English, it will be stored as one of the general rules, but the quality of the translations obtained using it will vary, depending on individual cases. This is illustrated in Fig. 6.5.

The transformations of the internal structure will differ depending upon the language into which the source text is to be translated. If we are interested in producing high-quality translations, then this part of the machine translation system must be developed in great detail. Indeed, it is often sufficient to know the capabilities of the transfer portion of the system to get an approximate idea of the quality of a machine translation system. Unfortunately, current linguistic theory teaches us very little about the nature of the transformations of the internal structure required, so that current systems have been constructed without

B having the property of A

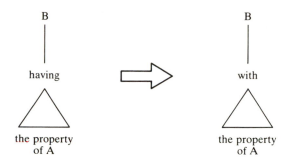

effect which is given to A

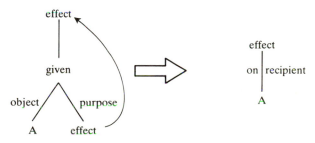

Fig. 6.5 Examples of application of the transformational rule shown in Fig. 6.4.

any underlying theoretical coherence in this regard. In truth, the problems of how a linguistic expression in one language will appear in a different language, and how situations and contexts will affect those expressions, lie at the point of contact between language and culture.

Comparative research in linguistics requires more than a thorough knowledge of two languages; it also requires the development of fundamental ideas and methodologies for rigorous comparison of languages. Further developments in this subfield of linguistics are therefore eagerly awaited. It is worth noting here that, quite aside from the practical applications of machine translation, such methodologies will undoubtedly be useful in comparative linguistic research. In other words, researchers in machine translation need not always seek the cooperation and research results of linguists, but may also be able to return the favour and provide insights into some of the issues of linguistics itself. Because linguistics is the literary discipline most closely linked to the natural sciences, exchange with the information sciences will undoubtedly become increasingly frequent. Only once close collaboration is achieved will truly sophisticated machine translation systems emerge.

Sentence generation

Once the internal structure of a sentence in one language has been transferred into that of a second language, the process of synthesis of the sentence in a readable form must begin. In many cases, the internal structure which is the starting point for sentence generation is expressed in the case grammatical terms. It can, however, be a complicated and troublesome process to generate sentences directly from the case structure representations, so that many systems first transform these into phrase structure representations and from there generate the final sentence in the target language. There are of course various forms of linguistic expressions to choose from: for example, the active or passive voice of the sentence, whether use of a noun phrase for a certain portion of the internal structure is suitable, or whether an embedded sentence structure should be used. Most of these structures are first changed into phrase structure representations and then treated within the framework of transformational grammar. This sequence of procedures is found to facilitate the translation process considerably. For example, an embedded sentence which acts as the subject of a relatively complex English sentence is often replaced by 'it' at the beginning of the sentence, and the embedded sentence itself is placed at the end:

It is true that he got the first prize.

In such cases, it is far easier to employ phrase structure grammar than try to deal with the sentence in the terms of case grammar internal structures.

The process of transformation from a case grammar to a phrase structure representation begins at the top node of the tree in the case grammar representation. Normally, the first node represents the main predicate of the entire sentence and, depending upon the kind of sentence structure required by that predicate, the positional relationships of the phrases for various case slots directly related to that predicate can be decided upon. Consider an example where the verb in the target language (English) requires an infinitive. If the noun in the source language is in the 'patient' case in the case grammar representation, the noun can often be rendered as the infinitive form of a verb. The appropriate phrase structure tree is then constructed. For example, in the Japanese sentence:

勝利をもくろむ

KATSURI WO MOKUROMU

victory is intending

'intend' might be chosen for the translation of the main verb MOKUROMU. Since 'intend' often demands a following infinitive form, we might choose a synonym for the literal translation of KATSURI (victory), such as 'win', and use a construction such as 'intend to win'. If there were an adjective describing 'victory', such as ZETTAI NO KATSURI [absolute victory], the noun KATSURI could be transformed into a verb and the adjective, ZETTAI NO, into an adverb, ZETTAI NI. In other words, 'absolute victory' would become 'win absolutely'.

In this way, the grammatical requirements which the first predicate in the case structure has on the various elements which follow it can alter the grammatical form of those elements. It is therefore necessary to undertake these procedures recursively, starting from the roots of the tree and proceeding to the ends of each branch, and to produce an appropriate tree for the resultant phrase structure.

Having achieved such a phrase structure, various other issues must then be addressed. These include decisions about certain prepositions, articles, and other words not found in Japanese but required in the English translation, the relationships between the predicate and its cases, and the relationships among the various nouns.

Japanese and English — some differences in sentence structure

It is often said that Japanese is a 'to be' language and English a 'to do' language. In other words, in Japanese, situational descriptions (centrally related to 'being') are frequently employed, whereas in English behavioural expressions (related to 'doing') are more frequently used. Both types of expression are of course available in both languages, but emphasis on one type or the other appears to have emerged.

In English, an expression such as 'The wind opened the door' is entirely acceptable, whereas in Japanese the literal equivalent is grammatically correct, but unlikely to be used:

風がドアを開けた。

KAZE GA DOA WO AKETA

wind door opened

The wind opened the door.

A better, less literal translation into Japanese would be:

風でドアが開いた。

KAZE DE DOA GA AITA

wind by means of door became opened

(the English translation of which would be 'The door was opened by the wind').

Similarly, a sentence such as:

地震で建物が壊れた。

JISHIN DE KENBUTSU GA KOWARETA

earthquake by means of buildings collapsed

would be analysed as having a structure such as shown in Fig. 6.6. If the noun phrase indicating causal potency is taken as the subject, the active verb corresponding to the passive verb ('collapse') would be 'destroy', and the structure of the entire sentence must be transformed as in Fig. 6.7. In the end, a translation such as the following would be obtained:

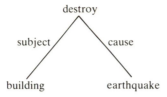

Fig. 6.6 The tree structure for 'The earthquake destroyed the buildings.'

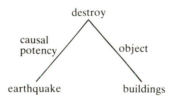

Fig. 6.7 An alternative tree structure for 'The earthquake destroyed the buildings.'

> The earthquake destroyed the buildings.

rather than the more literal translation:

> The buildings collapsed due to the earthquake.

Examples of this kind are numerous.

The subject (or agent) of a sentence is often omitted in Japanese. In such cases it is necessary to deduce what the subject is and to reconstruct it, but this often proves a difficult task for mechanical systems. Let us consider the possibility of constructing an English sentence which lacks an agent. The easiest example is the transformation of a sentence into the passive voice. The following Japanese sentence has no subject:

食べ物を与えた。

TABEMONO WO ATAETA

food giving was

There is no indication of who gave the food. The literal translation into English is therefore horrendous:

(   ) gave food (to ...)

but since this is not valid English, it is possible to change it into the passive voice:

Food was given.

By this means a valid sentence can be constructed without indication of the agent of the action.

Moreover, in Japanese, valid sentences can be constructed which have neither subject nor direct object. For example, in the following sentence:

実験と比較し、良い一致を見た。
JIKKEN TO HIKAKUSHI, YOI ITCHI WO MITA

experiment with compared, good agreement seen

the first half of the sentence does not indicate who did the comparing or what it was that was compared.

Let us see whether or not the predicate portion can be made into a subject in such a case. The noun form of the English verb corresponding to HIKAKU SHI ('to compare') would be 'comparison'. That noun can then be taken as the subject and a passive verb can be employed ('something was done'). A translation of the following kind is then obtained:

Comparison was made with experiment and good agreement was seen.

Some similar examples, which have been produced by our machine translation system at Kyoto University, are shown in Fig. 6.8.

It is worth noting that if a translation of these kinds of sentences were attempted using a pivot language method, the omitted phrases would need to be correctly inferred and some extremely troublesome processing would thus be required. In fact, it is found that the inferred phrases are not always correct using that technique, and mistranslations are frequently

LSI の発達でプリント配線基板１枚当りの
集積度が増大し、確実な試験法が重要になった。

LSI NO HATTATSU DE print HAISENKIBAN
ICHIMAI ATARI NO SHUSEKIDO GA
ZODAI SHI, KAKUJITSU NA SHIKENHO
GA JUYO NI NATTA.

LSI of development print circuit boards one per of integration
degree increase, reliable of testing method important became.

Advances of LSI increase the integration of printed circuit
boards per one piece, and reliable testing methods become
important.

沿面線上 streamer の伸展過程を駒撮り写真から検討し、
従来の沿面放電の伸展過程と比較

ENMENSENJO streamer NO SHINTEN KATEI
WO KOMADORI SHASHIN KARA KENTO SHI,
JURAI NO ENMENSEN HODEN NO SHINTEN
KATEI TO HIKAKU

surface wire on streamer of elongation processes high-speed
motion pictures from studied, conventional surface discharge of
elongation processes with comparison

Elongation processes of the streamer on surface wire are studied
according to high-speed motion pictures, and compared with
conventional development processes of the surface discharge.

ヘリウムで調べ次のことがわかった。

helium DE SHIRABE TSUGI NO KOTO GA
WAKATTA.

helium with examine next of facts understood.

The examination is made by the helium, and following facts
are found.

Fig. 6.8 Examples of machine translation results.

produced. In this sense as well, it must be concluded that, at least at the current level of technical capability, techniques employing pivot languages are not successful.

For this reason, the transfer technique is normally employed, but the transfer technique itself has the disadvantage of requiring an extremely large number of transformational rules. There is, moreover, the problem when using this method, of requiring a word dictionary with an extremely rich and informationally dense structure. A great deal of information for each term must be stored, and it must be stored in a form which makes it useful in a variety of contexts. If a word is, for example, a verb, its derivatives must be noted, its noun form recorded (often including several possibilities), as well as its adjectival form and its adverbial form. Moreover, its superordinate concept, synonyms and antonyms, and so on are also required.

If all of these procedures and all information of this kind are implemented in a machine translation system, quite sophisticated, human-like translations can be expected, but there remains the difficult problem of deciding what circumstances are appropriate for the application of these procedures and how thoroughly they should be applied in individual cases. It would be meaningless to insist that a predicate should be transformed into a subject in all cases, and even when both subject and object are omitted, there are frequently other appropriate expressions which can be used. For this reason, the decisions concerning when and where to use such methods will require further advances in research on context recognition and other higher-level aspects of language.

7 Machine translation systems

Actual procedures

In previous chapters, the mechanisms involved in machine translation have been explained, but when the entire system necessary for carrying out a translation by computer is considered, it is found that many additional elements are needed. First, once the original text has been provided, it must be decided if it is suitable for translation. Since most machine translation systems can only handle text from within a specific discipline, there is a need to see if the current text falls within that scope. Next, it must be considered whether the text is of an appropriate style. The third question concerns the purpose of the translation and who will read it.

In the case of the parliament of the European Community, where the text of a parliamentary speech is distributed in several languages, computer translation is virtually impossible: such papers are often brought to the human translators only hours before the speech is to take place, making for very tight deadlines. If the person who will read the translation is, for example, concerned with research of a technical-scientific nature, then it can be assumed that such specialists will understand the translation without difficulty: even if the translation is imperfect, the specialist's knowledge will supplement any deficiencies and thus enable understanding. Even in the case of a translation of a manual on the maintenance of a mechanical apparatus, the concrete nature of the topic and the fact that the reader will also be a specialist virtually guarantees that the reader will be able to comprehend the translated text. However, the common inversion of 'yes' and 'no' in translation or a mistake in the order of procedures to be undertaken has been known to have serious consequences; nothing can be left to chance. Such error in the sequence of steps to be taken in a repair manual could well result in loss of an expensive piece of equipment.

In 1982 an English–German machine translation system called

SYSTRAN was introduced into the European Community. The system requires that specialists be employed solely to decide if a given text is appropriate for machine translation. If so, the text is translated without human intervention. Together with various subsequent improvements in the system, it is said that the rate of successful translation has approached 90%.

The text to be translated has to be input into the system. Currently, this means manual input, usually typing, but when machine translation systems become more widely used it is likely that automatic text recognition systems will be used. The technology needed for reading both roman letters and the Japanese phonetic alphabet (*kana*) and ideographic characters (*kanji*) is already quite advanced, and sufficient accuracy for automatic text recognition of several styles of characters is now possible.

There is of course no technical problem involved in producing a printed output of the translated text. This is a well-known strength of modern computers. Currently, the typing and proofing of hand-translated text demands considerable time, effort, and cost. In the translation offices of the European Economic Community texts are translated and typed, then read and corrected by an expert in the field, and finally the text is retyped and sent for printing. Even with the advent of word processors, this process requires nearly two complete typings of the text, and the conventional process of printing also takes considerable time. If, however, some of these processes could be computerized, the cost in time and money could be greatly reduced. The first translation might be done by machine, followed by hand corrections made to the computer's output; then the entire formatting and printing process could be handled by computer. Even if the costs of correcting the computer translation remained high, this level of computer-assisted translation would nevertheless produce great economies.

One problem which arises in the input and output of text concerns the figures which are found in most non-fictional works. The explanatory terms which are often sprinkled throughout a figure must also be translated and placed in the new version. The actual translation of course causes no difficulties, but problems frequently arise in the placement of such terminology in the figures at the appropriate sites, particularly when the translation is between languages which do not read in the same direction.

The work involved in adjusting such figures is not conceptually difficult, but requires considerable time and indicates that machine translation systems will require quite sophisticated graphics capabilities, in addition to the translation procedure itself.

Software systems

Machine translation systems consist of several components (Fig. 7.1). In addition to the apparatus needed for text input and output, two important components are the translation machine itself and the system's dictionary. The translation machine is the software which undertakes the translation from one language to another, making use of the grammars needed (analytical grammar, transformational grammar, and generative grammar). In the older machine translation systems, no clear distinction was made

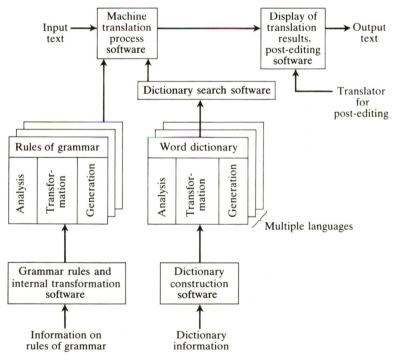

Fig. 7.1 The basic software of a machine translation system.

between the translation software itself and the set of grammatical rules, so that the rules of grammar were contained within the software program. Since, however, the grammatical rules must be updated regularly and a variety of languages handled, it is simply not practicable to make software changes for each and every language to be dealt with by the system. It is therefore necessary to construct software which has a basis common to all the languages handled by the system and into which the given language's grammar can be inserted.

When the design of the software system is poor, certain kinds of relatively low-quality translation may still be possible, but when attempts are made to improve the system so that it can handle more difficult texts and make more complex decisions, limitations will rapidly be reached. Ultimately, it will become impossible to improve the quality of the translations beyond a certain point.

Machine translation systems must be designed so that they can handle more than two specific languages — for example, Japanese and English, but also Russian and other languages which are based upon completely different structural principles. For this reason, systems are required which have the flexibility of being able to incorporate new linguistic principles and linguistic theories directly into their structure without a complete rebuilding of the system. Moreover, the systems must be structured such that the grammatical rules can be easily seen and the grammatical mechanisms easily understood. In other words, the software must be written such that not only computer specialists but also linguists can append or alter grammatical rules. This concept of flexibility is an absolutely fundamental consideration in the design of machine translation systems if they are to be upgradable, maintainable, and efficiently developed.

From the standpoint of software development, each step in machine translation can be viewed as a change in the tree structure. The grammatical rules can be seen as the transformation from tree structure$_1$ to tree structure$_2$. If a section of the given sentence has a structure corresponding to tree structure$_1$, then it will be automatically rewritten as tree structure$_2$. In general, the detection of a fit between the given sentence and a particular tree structure is called 'pattern matching'. Thus, a major effort in machine translation research has been made in the development of programming languages which perform such pattern matching as quickly as possible, using various algorithms for deciding on matching.

At the outset, of course, the given text is not in the form of a tree structure, but is merely a string of characters. Such a string, however, can itself be considered as a special kind of tree structure. First, by means of a dictionary search of character strings, morphemic units are detected and labelled as nodes (with indication of the appropriate part of speech attached). Next, several parts of speech are brought together and a partial tree structure is formed. This process is repeated until the entire internal structure of the sentence is determined. Only once the sentence has been fully analysed and transformed into its own internal structure is it then transformed into the internal structure of a foreign language. Thereafter, starting with the tree structure, it is gradually transformed into a character string and the translation produced.

Dictionaries

While sufficient organization using the grammatical rules is of course necessary, each word in a language has distinct properties of its own and strongly influences the shape of the sentence in which it appears. As a consequence, the degree to which the detailed description of each word's grammatical usage is contained in the system's dictionary will greatly affect the quality of the translation. Note that the set of specific usages of a given word is not the same as the generally applicable rules of grammar for the language, but comprises grammatical constructions which are specific for that word alone. For this reason, they are best stored in the computer memory as a part of the dictionary of words. The software system in machine translation must, therefore, first apply these rules of usage specific to the given words in the sentence, and only subsequently apply the more general rules of grammar.

Definite rules are yet to be established on how much information should be included in dictionaries. It is, however, now being realized internationally that as detailed a dictionary as possible must be used, and the details must be incorporated using a standardized format which virtually any kind of translation system can utilize. Having said this, it should also be noted that to construct a dictionary containing several tens of thousands of words with linguistically detailed and accurate information, an investment of time and money is required which is simply beyond the capacities of most private enterprises. For this reason, the involvement of many countries will be needed in the construction of high-quality dictionaries. This is the outstanding area where international cooperation will be essential.

For the actual process of translation, three varieties of dictionary are in fact required. These are the dictionary required for analysis of the input text, one which is used for the linguistic transformations from one language to another, and one for generating the translated text (Fig. 7.2). The analytic and generative dictionaries can in fact be one and the same for each language. For example, in translating from English to Japanese, in addition to the dictionary needed for translating between the languages, two dictionaries are required (one for analysis and one

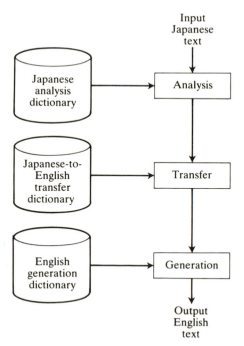

Fig. 7.2 Examples of machine translation dictionaries.

for synthesis), but they are the same two required in translating from Japanese to English.

One features of a dictionary which must be given constant attention during dictionary construction is maintenance of a consistent level of quality. The actual work of compiling a dictionary of such a size must of course be divided among many people, but the workers' abilities will vary. In the subtle areas requiring some degree of linguistic insight in particular, individual differences will inevitably be large. To prevent such differences from being reflected in the dictionary, a 'dictionary construction manual' must first be devised and the dictionary data then input by a team of workers all using that manual. Once the work of constructing the dictionary data has been completed, the data can be loaded into the computer and made use of in machine translation systems by means of appropriate 'dictionary utilization software' systems.

In the midst of experiments on machine translation, failures of translation are often discovered that are due to problems inherent to the data in the dictionary. It is thus absolutely essential that the machine translation system be constructed such that immediate dictionary revision is possible. At the same time, great caution is needed whenever alterations are being made to the dictionary data. When changes are made in order that a certain piece of text can be correctly translated, it may then be found that expressions which had previously been well translated can no longer be handled properly. As machine translation systems increase in size and complexity, it becomes more and more difficult to ascertain the influences which changes in one part of the dictionary will have on the overall system, and the time required to make even modest improvements increases dramatically as a consequence. It should also be said that there are cases when the performance of the entire system can be worsened, so alterations to the system should not be made lightly.

The words which are needed in a dictionary can be divided into two broad classes: a fundamental vocabulary of words which are needed in virtually all kinds of translation, and a list of specialized and technical terms needed for translation in a specific field. It is difficult to generalize about the number of specialized terms in any given field, but in many cases this number is several tens of thousands, and sometimes as large as a hundred thousand.

Needless to say, supplying a machine translation system with this many words is itself a mammoth undertaking. When a commercial machine translation system is delivered to an end-user, the fundamental vocabulary list is already supplied, but the specialized terms may or may not be included. In many cases, where the system is to be used within a narrow specialization, the user himself must supply some of the technical words and phrases. In general, the system will be constructed such that revisions of the fundamental vocabulary of the system by the user will not be allowed. This precaution is taken simply because imprudent alterations of the vocabulary would gradually make the system worse. Such systems are therefore built such that the user can freely append words to a portion of the dictionary which is left open by the developer. The user, not the maker, is thus responsible for the quality of this part of the dictionary.

Pre-editing and post-editing

Translation by computer is still an imperfect art, and may remain imperfect well into the future. Of course, this pessimism depends largely on what is meant by 'imperfect', and it should be said that high-quality translation by humans cannot be expected unless a highly experienced translator is employed.

Occasionally a novice translator will make the kinds of errors that are made by machines. In either case the text which emerges is frequently quite difficult to read as it stands — and usually includes mistranslations and omissions. For these reasons an ex-perienced translator must read the translation, correct mistakes, and produce more readable text. An example of such post-editing is shown in Fig. 7.3.

When experienced translators in the European Community have been given the output of machine translations (produced by SYSTRAN) for post-editing, it is often found that the meaning of sentences which have been poorly translated remains unclear even after several readings, and the human translators simply have no idea what kind of post-editing should be done. In such cases, the translators often feel that it would be simpler and easier to do the translation themselves, rather than post-edit a garbled machine translation.

It has therefore been concluded that, as a fundamental

米国ボストンの New England 電話会社において、マイクロ
コンピュー　タをベースとしたエネルギー管理制御システ
ムを設備し、エネルギー 原価の低減に成功した
実例を紹介。

The cases are introduced in which ~~the~~ success is made in the
reduction of energy costs by equipping the energy management and
control systems ~~based on~~ (with) microcomputers for (the) New England Telephone
Company ~~of the United States,~~ (located in) Boston.

また接合部の設計、位置の各種オプションについても紹介。

Then, each option of design and equipment of junctures is also
introduced.

製品から得られる各種ベネフィットの総和についての買者
の効用を測定する一つの方法にコンジョイント
測定法がある。

The conjoint measurement is ~~in~~ a method for measuring the utility
~~of buyers~~ (for buyers) about totals of each benefit which can be obtained from
products.

石油代替、省エネルギー機器の開発、代替エネルギー開発
の例、原子力発電、家電産業の動向についても解説。

Trends of substitutes for petroleum, examples of development ~~and~~
~~development~~ of alternative energy of energy saving devices,
nuclear power generation, and (the) household electrical industries are
also explained.

電荷の発生　その程度と保護対策

The generation of charges : degrees and protective measures.

Fig. 7.3 Examples of post-editing.

principle, machine tranlsations should be of such a quality that, at the very least, they do not cause confusion in those doing the post-editing. On the other hand, it is also found that when editors work with a large volume of machine-translated text, they gradually get accustomed to the quality of the computer output and get to know the patterns, idiosyncracies, and common mistakes in the translations. The post-editing process thus becomes easier with practice.

Post-editing requires human time and effort. Consideration must therefore be given to ways in which this work can be made easier and faster, and so less expensive. This reinforces the need for terminals with sophisticated on-line graphics display capabilities.

If the post-editing of machine-translated text is such difficult work, one can imagine that it would be beneficial if the original text were edited to produce sentences which would be easier for the machine translation system to handle in the first place. In other words, *pre-editing* might be a means for improving the quality of machine translations.

At the Xerox Corporation in America and General Motors in Canada, specialists for pre-editing are employed to write technical material according to pre-defined rules which will facilitate machine translation. Such rules include the use of a limited vocabulary, each word of which is used with a single meaning and can therefore be given a unique translation. Moreover, the structure of the sentences is maintained at a relatively simple level and conforms to a certain number of fixed patterns. Clearly, a certain period of training is required to be able to write along such predefined lines, but the result is that the original text can be given to a machine translation system and a translation produced which will require little or no post-editing.

In general, engineering systems are built on the basis of certain assumptions concerning their functional capabilities. For example, since it is presumed that most automobiles will be run on smooth roads, they are not designed to be capable of wandering through fields and over mountains. Robots also work within certain well-defined limitations; for example, they are capable of welding parts of specified sizes within a certain limited workspace. Since machine translation systems are also engineering systems, it is natural that they too should be designed with a first priority of being functional, and this may require certain presuppositions concerning their 'work space' — that is, the structure of the text to be translated. When texts which fall within the limitations of the machine translation system are provided, then a consistent level of quality in the end-product can be guaranteed, but it is sound engineering practice that when the limitations of the system are exceeded, no guarantee can be made concerning the final product.

In other words, if our intention is the construction of a trouble-free machine translation system, certain constraints concerning the nature of the linguistic expressions must be imposed. It must therefore be understood that, realistically, a general assertion concerning the quality of machine translations cannot be made, but rather that the quality of the translation can be assured only within certain limitations. To be sure, no one would say that automobiles are no good simply because they cannot be driven through swamps! Similarly, it would be a mistake to think that since machine translation systems can handle only a limited range of linguistic expressions, such systems are forcing human beings into the confines of a narrow realm of linguistic phenomena!

In order to evaluate machine translation systems from the point of view of their general applicability, a great many factors must be considered. The evaluation of the quality of such translations has already been discussed, but many other issues must also be dealt with, including the speed of the translation, the cost of translation (computer running costs, machine translation software costs, the costs of various dictionaries, etc.), the size of the staff needed for practical use, estimates concerning the cost of post-editing staff, the personnel required for maintaining the dictionaries, the approximate cost per word of enlarging the dictionary, and so on.

As discussed above, from the perspective of system development, the most important consideration must be that the system is designed such that changes and improvements can continually be made in the light of new linguistic theories and the continuing demands for ever larger dictionaries.

8 The future of machine translation systems

Towards higher-level translation capabilities

There are many linguistic phenomena which machine translation systems cannot yet deal with. Current systems can handle only single sentences at a time, and they function solely on the basis of the principle of compositionality. Thus, machine translations ignore the fact that correct translation often requires consideration of the relations between sentences.

Even at the level of the translation of single sentences, there are many difficulties which cannot yet be properly handled. For example, in the Japanese language, distinctions between singular and plural are often not made. In a literal translation of a commonly used construction, we might have:

TSUKUE NO UE NI HON GA ARU

desk(s) on book(s) is (are)

There is a (are) book(s) on (a/the) desk(s).

Since it is unclear whether there are one or several books and/or desks, an accurate translation simply cannot be made. Moreover, there is no appropriate translation for the English article 'the' in Japanese. Consequently, when translating Japanese into English there is no certain and easy formula for when 'the' should be used with a given noun, when 'a' should be used, when no article is needed, or when a plural is required. To a certain extent, these questions can be answered by linguistic theory and the relevant facts concerning such nouns included within the system's knowledge, but there are still too many cases when the answers are uncertain.

Negative expressions are also a problem. It is of course possible to translate:

彼は東京に行かなかった。

KARE WA TOKYO NI IKANAKATTTA

him as for Tokyo to there was not going

as 'He did not go to Tokyo', but the translation of:

彼は東京には行かなかった。

KARE WA TOKYO NI WA IKANAKATTA

him as for Tokyo to as for there was not going

is less unambiguous, in so far as the WA which follows the 'to Tokyo' phrase puts some emphasis on Tokyo. Possible translations include 'He did not go to Tokyo' and 'It was Tokyo to which he did not go.'

Another common problem concerns double negations. How should the double negation be handled in a sentence such as:

全く意味のない文書とは言えない。

MATTAKU IMI NO NAI BUNSHO TO WA IENAI

completely meaning having none sentence as for it cannot be said

In this example, the adverb MATTAKU could be a part of either negation. In one case, the sentence would be translated as:

It cannot be said that the sentence has no meaning at all.

and in the other it would be:

It absolutely cannot be said that the sentence has no meaning.

In general, there are two varieties of negation, partial and complete, which must be distinguished for accurate translation. When a partial negation is appropriate, it is necessary to determine the scope of the sentence which contains the negation. This is a task which computers find difficult to do.

Problems also arise with sentences such as the following:

Every man loves a woman.

人はそれぞれ好きな女性がいる。

HITO WA SOREZORE SUKI NA JOSEI GA IRU.

man as for respective love of woman is

すべての男が一人の女性を愛している。

SUBETE NO OTOKO GA HITORI NO JOSEI WO
AI SHITE IRU.

every of man one of woman loves

where the two translations distinguish between the possibilities of
all men loving one particular woman or each man loving a different woman.

In order to obtain accurate translations, not only must the
structure and meaning of individual sentences be carefully analysed, but the relationships with preceding and succeeding sentences must also be examined. The study of the connections and
relations between sentences is usually referred to as discourse
analysis or conversational grammar analysis.

In translating from Japanese to English, inferences concerning
abbreviated or omitted elements within the sentence is another
unavoidable problem. Certain aspects of making such inferences
have been successfully implemented in computer programs, but
there is no way of guaranteeing that the inferences are correct. In
any case, the inference mechanism is extremely complex and
requires considerable processing time, so that current commercial
machine translation systems do not normally include such
mechanisms. As discussed in Chapter 6, methods for searching
for linguistic expressions which can be understood regardless of
such omissions (and thus avoiding the problems involved) are
currently being pursued.

Even the translation of pronouns is not a simple matter. The
correct translation of the Japanese SORE (literally, 'that') is
often the plural, 'those', while the correct translation of the
English 'they' is normally KARERA when referring to people,
but SORERA when referring to things. Fortunately, such translations require only a small amount of contextual analysis. To a
certain extent they can be handled mechanically and have already
become a part of working systems.

One of the principal problems which must be considered in the
production of sentences from their internal representations is that
of the *focus* of the sentences. Linguists have previously explained

difficulties in determining the focus in Japanese as being due to the fact that the word sequence in Japanese is relatively flexible, but in fact this is wrong. On the contrary, the word-order within the sentence determines which parts of the sentence will receive emphasis. For example, if a sentence contains the following elements: a verb, 'to enter', a subject for the verb, 'he', and an object, 'Tokyo University', various sentences with distinctly different nuances can be produced, principally by altering the word-order.

彼は東大に合格した。

KARE WA TODAI NI GOKAKU SHITA

he University of Tokyo into was accepted

He was accepted into the University of Tokyo.

彼が合格したのは東大だ。

KARE GA GOKAKU SHITA NO WA TODAI DA

he was accepted the one as for the University of Tokyo it is

That into which he was accepted is the University of Tokyo.

東大に合格したのは彼だ。

TODAI NI GOKAKU SHITA NO WA KARE DA

The University of Tokyo into was accepted the one as for he is

He is the one who was accepted into the University of Tokyo.

In the second example, we already know that he entered somewhere, and the new information is the fact that the place is Tokyo University. In the third example, the new information concerns 'he'. These examples illustrate the fact that from the listener's point of view there is 'old information', either previously mentioned or presumed, and 'new information' which must be communicated in the sentence.

In Japanese, the new information usually comes immediately before the predicate, whereas in English the new information generally comes at the end of the sentence. For the purpose of translation, it is necessary to learn from the context provided which phrase in the sentence contains the new information (although in many cases this can be inferred from the word-order of the sentence). In other words, in order to choose the most appropriate translation from among several possibilities, it is essential to clarify what the new information in the sentence is.

A continuing problem in translation is to determine the degree to which a 'free' semantic translation should be made and, in the case of machine translation, to determine the degree to which semantic translation is even possible. In the latter case, it is virtually impossible to achieve a semantic translation which is totally unrelated to the structure of the original sentence. As discussed in Chapter 6, it is thought that machine translations can be done in which the phrases within a sentence and the structure of the sentence are structurally transformed, but completely semantic machine translation of literary sentences remains an impossibility.

Aside from the problem of such literary translation, numerous difficult problems remain. Current developments in theoretical linguistics promise to clarify many of the outstanding questions regarding the mechanisms of language, but it will be necessary not only to undertake theoretical linguistic research within given languages, but also to do comparative research explicitly for the purpose of learning the rules for translation between two specific languages. Finally, it will be necessary to conduct research in theoretical linguistics and cognitive science with the aim of eluci-dating the mechanisms used by translators as they undertake translation work. One problem of particular interest is the rela-tionship between what the translator wants to express and what linguistic expression is ultimately chosen. Even when what one wants to express is not stated explicitly, it is often implied some-where within the sentence. How can that implied meaning be discovered and made explicit for translation? This problem has hardly been touched upon in research on translation anywhere in the world, but it is a fascinating problem which lies at the heart of the mechanisms involved in human cognition.

The understanding of language

If the various problems discussed in the previous sections are to be solved, new directions must be pursued that reach beyond the limitations of traditional linguistic research. In other words, it is essential to clarify what it means for a human being to understand language, and then to consider how to build an electronic system which has similar functions. Individuals have massive stores of knowledge which they activate and use in the process of understanding language. Consider again the following two sentences:

> I bought a car with four doors.

> I bought a car with four dollars.

The difference between the sentence in which the referent for the prepositional phrase is 'car' and that in which it is 'bought' depends crucially upon the presence or absence of general knowledge concerning the fact that some cars have four doors.

It is evident that both for the understanding of language and for translation purposes general knowledge is essential. But these issues themselves involve problems concerning the ways in which such general knowledge is to be organized, stored, and activated. The fact that knowledge is required for language understanding was first convincingly demonstrated by Winograd in the early 1970s, and thereafter considerable research has been done on the problem of the computerized handling of knowledge, but very few clear-cut advances have been made. There have been successes in the organization of knowledge within a narrow technical field from a specific perspective, and its implementation in the form of expert systems. Such systems, however, have the disadvantage of becoming virtually powerless when the perspective is altered slightly or when the field of application is even slightly enlarged. The organization of knowledge in computer systems remains an extremely difficult problem.

Furthermore, not only is knowledge necessary for language understanding, but functions for making inferences based upon that knowledge are required. Consider, for example, the following sentence:

> One thousand grams of salt were dissolved in water, and then it was heated.

Here, 'it' does not refer to 'salt' or to 'water', but to water with salt dissolved in it, that is, the 'salt water' which was just made.

In other words, whatever the arguments about the grammaticality, people use sentences in which such a pronoun does not have an explicit referent which has previously appeared in the sentence or been described. Rather, the referent can be something which the entire sentence implies: a result which could emerge in the real world, given the facts in the sentence up to that point.

In order for such inferences about the real world to be made, one possible approach would be to implement within the computer a hypothetical representation of the events in the real world which are described by the sentence. That is, the world described in the text must be constructed in the computer, and there must be the capability to simulate the changes which occur according to the input language. Seen in this light, the sentence supplies the information to the computer and instructs it to undertake a particular kind of simulation. A complex process of this kind must be possible if flexible dialogue between man and machine is to be undertaken.

In practice, for a variety of situations the implementation of such simulations is not a simple matter, but such difficulties do not necessarily imply that machine translation is categorically impossible. As discussed in Chapter 3, provided that the scope of the text is limited to a specific topic or topics, there are a great many cases in which the gap between what the human being understands and what the computer is currently undertaking is not terribly large.

Let us take as an example a truly meaningless sentence:

I watched the green duty of a desk.

If told to translate it, most humans could produce a translation made up of appropriate words with appropriate word-order, even without understanding the sentence, simply by following the rules of syntax.

One reason why it has been difficult to introduce models of language understanding into machine translation systems is that, in language understanding models, only those sentences can be handled which are interpreted as valid in the light of previous knowledge about words already in the dictionary. If word relationships previously unknown to the system appear in the text to be translated, the system has no idea how to establish the correct semantic relations. In other words, the above sentence or 'the ocean is flying' or similar nonsensical sentences simply cannot be understood and therefore cannot be translated on the basis of understanding.

However, sentences are virtually always constructed to convey new information, and they attain their significance by doing so. By establishing new relationships among words, the sentence has its unique effect. This effect is achieved by means of syntax; even when the meaning is not clear, by following the grammatical construction of the sentence, a human can interpret the relationships among the words, come to an understanding, and accept the sentences as establishing a new relationship between already familiar words. At this level, at any rate, humans and machine translation systems can be seen as undertaking essentially the same kind of processing.

Looking back over the Japanese language translations which were done several decades ago on the works of the German Conceptual Philosophers, one cannot help but have doubts about whether the translators understood their subject matter. It is easy to criticize them simply by saying that the quality of the translations was poor, but there is a more constructive way of evaluating these translations. Taking the position that words are merely tools used for expressing thoughts and concepts, it can be argued that the contents of the thoughts themselves will be communicated in translation if the transformations are made properly at the level of the tools (words) — quite aside from questions concerning the thoughts themselves. In other words, if the linguistic relations in the Japanese correspond to the same relations in the German original, then whatever meaning is contained in the translation reflects that present in the original.

If we consider the situation from the reader's perspective, then it is clearly desirable to have sentences which are written explicitly to convey unambiguous ideas. Whether or not the contents will be correctly understood will then depend upon the reader himself. In this sense, it can be said that Montague's semantics, where one is concerned with the true spirit of what a sentence means, is inappropriate for the purposes of machine translation.

Rather than argue that machine translation is impossible if one does not enter the world of language understanding (which is a virtually endless task), it may be wiser to search for a basis on which automatic translation is possible without getting heavily involved in issues of language understanding. Even if some corners must be cut, ways might be devised such that machine translation can be of practical use.

The future of practical systems

The debate whether or not machine translation is possible can be seen as a question of what range of linguistic expressions can be treated by computer and, within that range, what the situation is in practice. It is simply not a serious argument to maintain that since machines cannot breezily perform literary translations such as those done by the best human translators, therefore machine translation is impossible or useless. A few of the fields in which machine translation might be put to excellent use are listed below, but there are undoubtedly numerous other applications which are yet to be considered.

Large-quantity, high-speed translation

As discussed in detail above, the translation of a huge volume of technical material for industry is typical of the large quantity and high-speed translations which are needed. A huge volume of technical and scientific information continues to accumulate, so that a machine translation system in conjunction with an information retrieval system could be used for searching through a mammoth, truly international information database. Such a system would greatly facilitate the presentation of information to colleagues in similar academic fields, but with different native tongues throughout the world. Since the majority of Japanese technical and scientific information is written in Japanese, there is mutual dissatisfaction with the fact that information cannot be obtained from abroad, and the translated presentation of such information would be welcomed by all concerned. It goes without saying that the Japanese contribution in many fields would be greatly increased if the scientific and technical information and copyright materials of the Japanese Information Center for Science and Technology (JICST) or the Copyright Information Center could be translated by machine into English and other languages for presentation abroad.

Similarly, it would be extremely convenient if the news items sent out by the news services throughout the world and the telex dispatches of commercial enterprises could be quickly translated into and out of Japanese and distributed in the translated form. Although many of Japan's foreign correspondents for the news media and business can speak English, they have to deal with a large volume of information and must determine quickly what information is salient. Working in one's own mother tongue would undoubtedly be beneficial and increase efficiency several-fold. From this perspective, it is clear that a very high quality of translation may not be essential; it may suffice to have the raw output from the machine translation without the additional polishing of human post-editing. If one then scans such sentences of mixed quality, but in one's native language, and discovers information which appears to be of some importance, then that article alone could be translated by a professional translator or post-editing done of the machine translation output.

Small-quantity, private translations

The importance of translations for private use should not be overlooked. Consider, for example, the case where a certain text needs to be translated into French or Russian for use in a letter, but one has absolutely no knowledge of the target language. In such a case, one could rely upon an experienced translator or a machine translation, but clearly an awkward machine translation would not be satisfactory. If, however, an interactive machine translation system could then be consulted, many of the potential problems in a machine translation could be avoided. Such a system could be given the text for analysis and if ambiguities remain, the system itself could interrogate the user interactively. Since in this kind of scenario there is no problem with regard to the efficient use of time, translation of any desired quality could be obtained.

Questions from the machine translation system to the user would be of several varieties. The system would demand clarification from the user when, for example, a sentence is too long to be analysed, when a sentence contains a semantic ambiguity, when one of several meanings of a term needs to be confirmed, and so on. This kind of interactive system would ask questions confined to the scope of the original text (without real world knowledge) and would be designed solely to clarify the intentions of the author prior to the automatic translation procedure.

Undoubtedly this kind of personal use of machine translation systems will increase in the future. Once they have become available on personal computers, such letters could be translated using direct links between personal computers which have the appropriate translation capabilities at either end.

Machine translation applications

A variety of applied systems making use of machine translation can easily be envisaged. One particularly useful example would be their inclusion in language learning facilities. A more ambitious application would be the development of an automatic telephone interpreting system. One could then speak Japanese into such a telephone and have English or French emerge at the receiving end, or vice versa. If such a system could be realized, it would, at the very least, become possible to make routine enquiries and reservations or place orders virtually anywhere in the world as easily as is now possible within national boundaries. In such a machine interpreting system, one could obviously not simply chat in as relaxed a manner as with a friend, but would need to speak with clarity and use simple language. When a sentence could not be automatically interpreted, the system would be designed to request restatement of the sentence in a more simplified form. With such a system, results comparable to those obtained through machine translation should be possible.

The chief obstacle to a functional automatic interpreting system, however, lies not in the linguistic problems which face machine translation, but in the voice recognition capabilities which would need to work for any individual using the system. This is a formidable problem, but it might be solved within a decade or two in the following way. Each individual wanting to use the automatic interpreting system would be given a digital card which contains his fundamental voice parameters. These parameters would be permanently recorded on the card following a voice-recognition session during which the user spoke several thousand words slowly and distinctly into the system. When actually making use of the interpreting telephone, one would need only to insert the card into the telephone, where the voice parameters would be temporarily stored and the system adjusted to the idiosyncracies of the individual's voice for private use.

Needless to say, further research developments along these lines are still needed, but the directions for further research are already clear.

The need for long-term basic research

It is often said that, regardless of the area of application, the time span from the start of basic research until full realization of a useful product is on the order of 30 to 50 years. More than 25 years have elapsed from the start of machine translation research around 1960, and the emergence of practicable and economical systems is just about in sight. The fact that, in comparison with a host of other engineering, and specifically computer-based, systems, a relatively long period has been required for realization is due primarily to two factors. The first is quite simply that language is an expression of the highest intellectual functions of man. Secondly, there was a break in research activity for more than ten years, starting in the mid-1960s. Even supposing that we are now on the threshold of practical applications of machine translation, another ten or fifteen years will be needed before machine translation systems are put to use throughout society. Machine translation is still little more than a new-born child, and there are still many difficult lessons to be learned before we have a mature technology at our disposal.

It must therefore be asked what can be done to ensure the steady maturation of the art and science of machine translation. In a word, the answer is to conduct patient and prolonged research on each of the many aspects of translation. In Europe and America, there are today universities where translators are trained and the methods of translation, problems of vocabularies, and related issues are researched, both theoretically and practically. In Japan, translation is taught at only a handful of specialized schools and there is currently not a single university which has such a department. There is consequently a great need, particularly in Japan, to conduct linguistic research using the methods of the natural sciences. Since languages are continually changing, unexpected problems will probably always appear in machine translation, but for this very reason it is essential that continued theoretical and applied engineering research in linguistics be done.

Currently, machine translation is being pursued among the

principal languages of the world, but in the future machine translation systems will undoubtedly be developed for virtually all of the known written languages. It can also be said that machine translation may be particularly important for those languages where the native population is relatively small. In Japan today there is a considerable number of people involved in the translation of texts into and out of English, but what about translations of Arabic or Swahili? Despite the fact that a huge number of people speak those languages, there are certainly no more than a handful of Japanese who do so. Without exception, they are university teachers or employed in specialist work of some kind and therefore simply cannot give the time and energy required for the translation of a large volume of technical material. We must therefore think along the lines of having specialists in such languages participate in the development of machine translation systems, which can then be used for the translation of large volumes of material.

Machine translation is a difficult technological challenge, but if these issues are addressed squarely, capable systems can certainly be developed. As a by-product of such research, a wide technical base will be developed and numerous marketable applications will emerge. Moreover, such research on language is an important part of the elucidation of the mechanisms of the human mind and brain and, in many respects, is an inherently fascinating topic to pursue.

References

1. *Research on mechanical translation*, Report of the Committee on Science and Astronautics, US House of Representatives, Eighty-Sixth Congress, June 25, 1960, Union Calendar No. 895.
2. *Mechanical translation*, the journal on machine translation published between 1954 and 1964, edited by W.N. Locke and V. Yngve and published at MIT.
3. Starting in 1960, many Japanese reviews of the state of the art in machine translation can be found. The earliest were papers by H. Wada: 'A non-calculating calculator', *Information processing*, Vol. 1, 1960; 'Machine translation', *Information Processing*, Vol. 3, 1962 (Jpn.). The history of machine translation and the current state of machine translation in Japan are discussed in a special volume of *Information Processing*, Vol. 26, 1985 (Jpn.).
4. *Language and machines: computers in translation and linguistics*, National Academy of Sciences/National Research Council (1960) (generally known as the *ALPAC Report*).
5. N. Chomsky. *Syntactic structures*. Mouton (1957).
6. N. Chomsky. *Aspects of the theory of syntax*. MIT Press (1965).
7. C.J. Fillmore. 'The case for case'. *In Universals in linguistics theory* (E. Bach and T. Harms, eds.). Holt, Rinehart and Winston (1968).
8. Association of Japanese Electronics Industries. 'A survey of machine translation systems', 57-C-438 (March, 1982), 58-C-493 (March, 1983), 59-C-496 (March, 1984).
9. T. Kunihiro. 'A comparison of the structure of vocabularies'. In *Meaning and vocabulary*. (1981) p. 36.
10. C. Hockett. 'Grammar for the hearer'. In *Structure of language and its mathematical aspects*. (Proceedings of a symposium in applied mathematics, Vol. XII), American Mathematical Society (1961).

Some representative books on machine translation.

B. Henisz-Dostert, R.R. MacDonald, and M. Zarechnak. *Machine translation*. Mouton (1979).

B.M. Snell (ed.). *Translating and the computer*. North-Holland (1979).

V. Lawson (ed.). *Practical experience of machine translation*. North-Holland (1982).

W.J. Hutchins. *Machine translation: past, present, future*. Ellis Horwood, 1986.

S. Nirenberg (ed.). *Machine translation: theoretical and methodological issues*. Cambridge University Press, 1987.

Montague, R.M. (1974) *Formal Philosophy*, ed. R.H. Thomason, New Haven, Conn.: Yale University Press.

Index